The American and Japanese
Auto Industries in Transition

The American and Japanese Auto Industries in Transition

Report of the Joint U.S.-Japan Automotive Study

edited by

Robert E. Cole
and
Taizo Yakushiji

conducted under the general direction
of research chairmen

Paul W. McCracken
and
Keichi Oshima

Ann Arbor

Center for Japanese Studies
The University of Michigan

Tokyo

Technova Inc.

1984

*Open access edition funded by the National Endowment for the Humanities/
Andrew W. Mellon Foundation Humanities Open Book Program.*

Copyright© 1984

Center for Japanese Studies
The University of Michigan
108 Lane Hall
Ann Arbor, MI 48109

and

Technova Inc.
13th Floor, Fukoku Seimei Building
2-2 Uchisaiwai-cho, 2-chome
Chiyoda-ku, Tokyo 100
Japan

Cover Design by Carrie Di Lorenzo
Calligraphy by Shunichi Kato

Library of Congress Cataloging in Publication Data

Joint U.S.-Japan Automotive Study.
 The American and Japanese auto industries in transition.
 Includes bibliographical references.
 1. Automobile industry and trade — United States.
2. Automobile industry and trade — Japan. I. Cole, Robert E.
II. Yakushiji, Taizo, 1944- . III. Title.
HD9710.U52J64 1984 338.4'76292'0952 84-5814
ISBN 0-939512-20-3

Printed and bound by CPI Group (UK) Ltd, Croydon, CR0 4YY

ISBN 978-0-939512-28-7 (paper)
ISBN 978-0-472-88001-0 (ebook)
ISBN 978-0-472-90205-7 (open access)

Contents

Contents

Figures

Figures

Tables

Foreword

In the Spring of 1980, the prospects for the U.S. auto industry stood at low ebb, and there were rapidly growing trade frictions between Japan and the United States. At that time, a number of us initiated a set of private discussions that led to the formal establishment of the Joint U.S.-Japan Automotive Study in September 1980.

At the time of this writing, conditions have once again changed, and the sighs of collective relief can be heard widely throughout the respective auto industries. That the situation changed so rapidly in such a short time, however, should make us mindful that the future holds little promise of tranquility or ease of predictability.

Paul W. McCracken
Keichi Oshima
Research Chairmen

Preface

This report was prepared for the Policy Board by the U.S. and Japanese research staffs of the Joint U.S.-Japan Automotive Study under the general direction of Professors Paul W. McCracken and Keichi Oshima, with research operations organized and coordinated by Robert E. Cole on the U.S. side, in close communication with Taizo Yakushiji on the Japanese side. The range of views within the Policy Board, representing widely varying backgrounds, perspectives, and interests, is great. The report is not what any one member of the Policy Board would have written. Therefore, each member reserves the right to his own views on specific matters. The sponsors listed on a subsequent page have joined the study with the intent to promote understanding and open dialogue on the subject of the U.S. and Japanese automotive industries. The report, while taking sponsors' views into account, does not necessarily reflect their individual or organizational views of issues covered in the report. Finally, individual researchers associated with the project do not themselves necessarily agree with all statements in the report. Notwithstanding, there was a substantial measure of consensus within the Policy Board and among the research scholars as to the nature of the problems and policy options of the auto industry. Therefore, we commend this report, produced by scholars in both countries, to the general public.

The Policy Board of the
Joint U.S.-Japan Automotive Study

Policy Board Members

Research Scholars

George BITTLINGMAYER
Assistant Professor of Business
  Administration
The University of Michigan

John CAMPBELL
Director
Center for Japanese Studies
The University of Michigan

David E. COLE
Director
Office for the Study of Automotive
  Transportation
The University of Michigan

Robert E. COLE
Professor of Sociology
The University of Michigan

Michael S. FLYNN
Associate Research Scientist
Joint U.S.-Japan Automotive
  Study
The University of Michigan

Mieko HANAEDA
Project Researcher
Technova Inc.

Lawrence. T. HARBECK
Associate Research Scientist
Office for the Study of Automotive
  Transportation
The University of Michigan

Richard P. HERVEY
Management Consultant and
  President
Sigma Associates

Masakazu IGUCHI
Professor of Mechanical
  Engineering for Production
The University of Tokyo

Motoo KAJI
Professor of Economics
The University of Tokyo

Fumio KODAMA
Associate Professor of System
  Engineering
The Saitama University

Jeffrey LIKER
Assistant Professor of Industrial
  and Operations Engineering
The University of Michigan

Paul W. McCRACKEN
Edmund Ezra Day Distinguished
  University Professor of
  Business Administration
The University of Michigan

Keichi OSHIMA
Professor Emeritus of Engineering
The University of Tokyo

Vladimir PUCIK
Assistant Professor of Business
    Administration
The University of Michigan

Haruo SHIMADA
Professor of Economics
Keio University

Robert J. THOMAS
Assistant Professor of Sociology
The University of Michigan

Adrian TSCHOEGL
Assistant Professor of Business
    Administration
The University of Michigan

Richard C. WILSON
Professor of Industrial Engineering
The University of Michigan

Taizo YAKUSHIJI
Associate Professor of Political
    Science
The Saitama University

Tsutomu YASUI
Manager, Overseas Operations
Technova Inc.

Policy Board Recommendations

The Policy Board members of the Joint U.S.-Japan Automotive Study, having reviewed and discussed the final report by the U.S. and Japanese research staff, come to the following conclusions. We are greatly concerned about:

— the maintenance of the liberal international economic order as a key to consumer welfare and world economic prosperity;
— the painful economic dislocations, in particular, high unemployment, that have occurred in recent years;
— the establishment of stable, long-term, competitive economic relations between the United States and Japan, especially in recognition of the increasing disturbances to our political relationships caused by economic disputes; and
— the maintenance of a prosperous and competitive automotive industry worldwide.

The Policy Board members recognize that:

— Automobiles and trucks will have a central role in meeting personal transportation needs in the future, even though the overall rate of growth may be slower than in the past.
— The automotive industry will evolve rapidly in the future, an evolution heavily influenced by the changing concept of a motor vehicle in response to changing consumer aspirations and demands.
— The driving forces for this stepped-up transformation of the industry, in particular, are the renewed efforts to develop new process and product technology and the increasing internationalization of the industry. These two factors are inextricably linked, and the way in which they combine and evolve will be a major force in shaping the future character of the automotive industry worldwide.
— The U.S.-Japan automotive issues have to be dealt with in ways consistent with the joint prosperity of both countries.

xix

The Policy Board members therefore recommend that:

1. The Governments of the United States and Japan should pursue trade policies, exchange-rate relationships, administrative policies, and monetary and fiscal policies that minimize market distortions.

2. The Japanese automotive industry, in recognition of the international environment and the impacts of its sales in individual countries on employment and economic dislocation, should continue to expand its commitment to local production where feasible or consider other alternatives to deal with these impacts.

3. The U.S. automotive industry should continue and intensify its efforts to enhance its international competitiveness by narrowing the gap in production costs and improving quality, strengthening its ties with parts and material suppliers, and developing new and more cooperative labor-management relations.

4. The automotive industries in both countries should enhance their capabilities, including their capital and human resources, so that they can be responsive to technological developments and changing consumer demands as the functions required of a motor vehicle continue to evolve.

In conclusion, the Policy Board members caution that the governments and automotive industries of both countries should maintain and intensify their efforts toward those objectives embodied in these recommendations, especially when improving economic times may mask underlying problems.

Executive Summary

Japan and the United States are the two most powerful industrial nations in the free world, and each is the other's largest overseas trading partner. This relationship has contributed substantially to the last thirty years' expansion of the world economy and a more liberal international economic order. But when the weakness of the world economy became sharply accentuated after 1979 and the trade imbalance increased, trade frictions between the two countries became serious. The automotive sector produces the most visible case in point.

To be sure, the recovery of the U.S. economy that began early in 1983 is encouraging, but the problems of specific sectors, such as the auto industry, are not thereby automatically solved. The auto issue between Japan and the United States is representative of the much broader problem of the economic relationship among the industrialized nations.

In view of the importance of stable, long-term economic relationships between Japan and the United States, automotive issues have to be dealt with in ways consistent with the joint prosperity of both countries. Furthermore, the current economic friction has the potential to adversely affect future political relationships. Indeed, under conditions of economic stagnation, major economic issues inevitably become political issues.

With these considerations in mind, the Joint U.S.-Japan Automotive Study project was started in September 1981 to determine the conditions that will allow for the prosperous coexistence of the respective automobile industries. During this two-year study, we have identified four driving forces that will play a major role in determining the future course of the automotive industry of both countries. These are: (1) consumers' demands and aspirations vis-à-vis automobiles; (2) flexible manufacturing systems (FMS); (3) rapidly evolving technology; and (4) the internationalization of the automotive industry.

As a product subject to the tastes and aspirations of the mass public, the automobile is changing its shape and functions according to rapidly evolving consumer preferences. In order to better satisfy the increasingly varied and rapidly changing consumer tastes, flexible manufacturing systems are developing—both strategically and in the specific sense of manufacturing technology. The popular fascination with robotics is misleading in that robotics constitute but one small manifestation of this larger process that is impacting the very organization of factories, the work process, manufacturer-supplier relationships, product-development strategies, and marketing practices. Most profoundly, rapid technological innovation is occurring within and along the periphery of the automobile industry.

xxi

The quickening pace of technological innovation is not limited to the highly publicized area of electronics but spreads into other major areas such as new materials and process technology. Again, these developments are of sufficient magnitude that they are disrupting established production practices, product-development cycles, marketing and distribution systems, capital-equipment investment plans, and organizational structures. They also challenge—and in many cases make obsolete—existing managerial competence, work practices, and skills of employees.

Lastly, rapid internationalization of the auto industry in all its varied aspects is transforming the industry. Cooperative arrangements among the producers and worldwide sourcing of parts and components are but two of the most visible aspects of this fundamental alteration of the industry. Issues of location of production from an economic and political perspective are being brought into sharper focus.

Furthermore, these four driving forces interact in a variety of ways and, in so doing, are rapidly changing the basis for competition among firms. Existing competitive relationships are being heightened. At the same time, specific cooperative relationships among producers are growing under the pressure of the huge investments required for firms to remain competitive and to provide sufficient return on these investments. The stakes are high for employees, consumers, shareholders, and the governments that benefit from the location of automotive companies within their national borders. In this report, we outline the workings of these processes as they impact in myriad fashions upon the industry.

These dramatic internal transformations of the industry are occurring in the context of crucial macro political and economic problems, including the recent, deep worldwide recession, large and rapidly shifting trade imbalances, sharp fluctuations in exchange rates, and high unemployment levels. This has inevitably led to a politicization of the issues. While various *ad hoc* measures to accommodate to these pressures are and will be taken, they have the potential to threaten severely the ongoing processes of technological innovation, as well as to impair many of the other changes we have outlined. To allow this to happen would work against providing consumers with products that serve their increasingly diverse preferences. The core of the dilemma facing the industrialized nations is how to make short-term adjustments to the macro political and economic problems without succumbing to a deteriorating cycle of long-term restrictions, stagnation, and decline.

With the objective of moving toward resolution of this dilemma, as well as furthering the adjustments to the new competitive forces we have outlined above, our study narrowed the problems to the following eight areas: (1) macroeconomic policy; (2) exchange rate; (3) market access; (4) technological progress; (5) manufacturing cost differences; (6) manufacturer-supplier relations; (7) human resource management; and (8) public policy. These eight areas interact with the four driving forces in critical ways that will determine the shape of the automotive industry of the future. In particular, the demand factor serves as a background condition that impacts all of the areas; similarly, the process of internationalization leads to changes in all areas. Our conclusions from studying these issues are as follows:

Macroeconomic Policy

The fundamental requirement for a strong auto industry is that governments of the industrial nations manage economic policies in a way that enables economic activity to expand along an orderly path and rapidly enough to steadily reduce currently high levels of unemployment in the entire industrial world. Specific "auto policies" cannot offset the deficiencies of basic economic policies that produce chronic stagnation. Conversely, effective performance by auto producers will not produce the desired outcome if government does not create a favorable environment. When protectionist measures are deemed necessary, they should contain an explicit schedule for their dismantlement in order to insure that the industry accepts the inevitability of the need to adjust to competitive forces. Experience in other industries has shown that once a managed trade regime is instituted, it is difficult to terminate.

Exchange Rate

The governments of the major developed economies should further coordinate their basic monetary and fiscal policies to avoid wide swings in exchange rates, which result from responses to large financial flows and to the "overshooting" phenomenon. A persistent maladjustment in the level of rates is the most fundamental trade problem. The Japanese government should continue its policy of financial liberalization, including the freeing of interest rates from administrative control and the opening of Japanese capital markets to foreigners, as well as providing greater access for Japanese institutions—all looking toward the more extensive use of the yen as a major international currency.

Market Access

Whereas barriers to the Japanese automobile market were once very high, this study stresses that the Japanese have reduced import barriers and that American automotive producers and public officials should recognize this fact. They need to further recognize that, increasingly, the major barriers to participating in the Japanese market are competitive costs, quality-oriented

Japanese customers, and service facilities. Furthermore, the building of strong dealer networks is critical to the success of such efforts.

At the same time, as part of its continuing efforts, the Japanese government should remove the last vestiges of protectionism in the automotive sector. This includes "leaning over backwards" to modify normal commercial practices to encourage imports. It is expected that emphasis will be shifted from barrier reduction to import promotion; an important symbol would be a decision by the Japanese government to add foreign-produced cars to their procurement activities.

Yet, even with all conceivable trade barriers removed, it is unlikely that there would be a significant increase in U.S. automotive exports to Japan. Transportation costs, development costs for products suitable to the Japanese market, the costs of producing for the Japanese market, the existence of productivity differentials of some magnitude, and lower labor costs make it clear that American cars are not competitive in the Japanese domestic mass market for the forseeable future.

Technological Progress

Regarding actions relating directly to the competitive bases of the industry, this study stresses the need for producers to enhance their technical capabilities to meet new market demands. It is our judgment that the possession of these technical capabilities and the ability to translate them into competitive products will increase the probability that existing firms will survive and grow in the future. This, in turn, suggests the need for expanded research and development activities, strong capital-investment programs, flexible manufacturing systems, and increased attention to manpower recruitment, training, and retention in technical areas.

Manufacturing Cost Differences

The U.S. automotive industry currently faces a real and substantial disadvantage in manufacturing costs compared to the Japanese industry. The significance of this difference is influenced by many factors, but it is clear that it constitutes a long-term problem that must be addressed. Significant progress in cost reduction has already been made by U.S. producers, but the results have been offset, to some degree, by a further strengthening of the dollar against the yen (heavily influenced, as mentioned earlier, by financial flows). In view of this current situation, some researchers emphasize that, in its marketing strategies in North America, the Japanese industry must be cautious in taking advantage of its manufacturing cost difference. A lack of caution could give rise to irresistible political pressures on the U.S. and Canadian governments to severely and permanently restrict Japanese access to North American markets. Other researchers stress that such restraint is a restrictive practice that would sacrifice consumer interests and that, therefore, the only solution is for the U.S. industry to reduce the manufacturing cost difference in a concerted and consistent fashion through a long-term strategy.

Manufacturer-Supplier Relations

Of overriding importance is the need for U.S. manufacturers to develop more cooperative relationships with their suppliers in moving toward joint efforts at technological innovation, product design, higher quality, and reduced costs. Many such efforts are under way, and they need to be intensified. On the other hand, it is important for Japanese producers to move toward a more internationally based supplier network. This represents one facet of the auto-industry internationalization that is stressed in this report.

Human Resource Management

Our analysis of human resource development as a management strategy highlights its critical importance to success. Fostering a productive, creative, and adaptable work force can be one of the single most important measures that a firm can undertake over the long term to improve its competitive standing. In addition to the obvious need for labor-management cooperation to bring about this objective, information sharing between management and employees has been identified as a critical element in this strategy. Properly done, information sharing can build trust and serve as a resource for joint problem solving. Japanese producers over the years have developed quite effective policies in these areas. We are encouraged by the recent efforts of American firms in this area as well, and we urge an intensification of these efforts. Some recent signs of "backsliding" in U.S. labor-management cooperation make it clear, however, that continuous efforts to sustain such activities are necessary.

Public Policy

Finally, with regard to public policy, we conclude that public policy, though important, has not been the major determinant of the success or problems of the auto industry in either the United States or Japan. Moreover, contrary to frequently encountered generalizations, governmental policy in both countries has produced mixed effects, aiding the development of the automobile industry in some areas and periods, hindering it in others.

In the case of Japan, during the critical "infant-industry" period of the 1950s, the auto firms benefited from a variety of government supports: long-term loans, accelerated depreciation, tariff exceptions on needed production machinery, tax breaks for export promotion, and others. With some exceptions, these were neither crucially important nor specifically designed for automobiles. The key government policy was thoroughgoing protection against both imported cars and, particularly, direct foreign investment.   By the mid-1960s Japanese government officials became increasingly aware of the problems this created as the strength of the Japanese auto industry grew. The 1970s saw a gradual dismantling of these barriers.

Currently, there is much discussion in the United States as to the character of Japanese industrial policy and its applicability to the United States. Our investigations of the automobile industry suggest that the key lessons to be learned are increased communication between industry and government, joint discussions about current trends and future problems, and a long-term focus on the issue of international competitiveness. In the case of the United States, governmental efforts to improve the international competitiveness of American industry cannot be continually subordinated to concerns about domestic or defense policy without serious long-term damage being inflicted on industry and those who depend on it for their livelihoods.   In this connection, it is clear that many currently unemployed automobile workers are unlikely to be rehired. Public policy to aid reemployment should be examined, as should regional policy to redevelop areas dependent on the automotive industry.   Both Europe and Japan provide several models of public-private cooperation in such programs that merit serious investigation.

Chapter 1
Introduction

Objective and Background of the Study

When the weakness of the world economy became sharply accentuated after 1979 and trade imbalances increased, trade frictions between Japan and the U.S. became serious. Tension over these matters still persists in spite of the three-year, voluntary trade-restraint measures and the one-year extension that resulted from the political judgments of both governments. These judgments were based on the poor state of the U.S. economy and, in particular, unemployment problems. They represented a "voluntary action" taken by the Japanese to minimize protectionist pressures in the United States and to give the U.S. automakers time to regain their competitive strength and readjust to changed market conditions.

The recovery of the U.S. economy that began early in 1983 is encouraging, but the auto industry's problems have not been automatically resolved. The auto issue between Japan and the United States is rooted in the more fundamental problem of the economic relationship among the industrialized nations, the origins of which can be traced, to a great extent, to recent changes in the world economic climate—such as the volatility and uncertainty of energy prices, the changing availability of natural resources to specific countries as a result of rapidly shifting prices, sharp fluctuations and imbalances in exchange rates, decreases in economic growth rates, and the changing balance in the relative economic strength of the industrialized nations. As such, they apply with equal force to Western Europe and have a major impact on the developing nations.

Governmental restrictions in such a key area as the automobile industry threaten the viability of the liberal international trading system. Yet, at the same time, we cannot ignore the central role of the automobile industry in many national economies, the employment impacts that result from large and sudden shifts in trading patterns, and the industry's significance for national security. Under conditions of economic stagnation, tremendous political pressure builds up to redress imbalances, often in a bilateral fashion. This is especially the case because these seemingly "automotive" issues are not only a matter of one specific industrial sector. They involve several major sectors, such as the steel and machine-tool industries, that are important for the whole economy of the respective nations. How should we respond to political pressures for a resolution of these economic problems? What conditions will allow for the prosperous coexistence of the respective automobile industries?

1

Framework of the Study

In the course of this study, we identified several critical issues. One concerns macro political and economic problems such as trade friction, unemployment, and decline in economic growth rates. While recognizing that these are indispensable concerns that establish critical parameters of what is possible, this study focuses on the evolution of the automotive industry both in an aggregate sense and in the paths to be taken by firms and national auto industries. In particular, we investigated the specific characteristics of auto firms and national auto industries that can lead them to a renewal in the future.

The term "renewal," of course, has different meanings for different companies and the different national industries. In the case of the United States, renewal generally involves regaining a competitive edge, lost as a result of the shift away from its unique large-car market to a situation where its markets are exposed to highly efficient worldwide competitors. In the case of the Japanese auto industry, the issue concerns how to restore a growth trajectory for the industry now that its domestic market has become largely a replacement market and significant worldwide restrictions on Japanese car exports have been instituted.

There are those who argue that the automobile industry will take the same path as that of the iron and steel industry or the shipbuilding industry. That is, when the world market starts to be saturated by expanded production capacity and enhanced productivity, an industry that fails to introduce a new technological breakthrough can be characterized as highly mature and subject to future decline. This suggests, of course, that technology is one of the driving forces for market growth and, thus, that the auto industry should be concerned with the conditions that foster technological innovation.

With regard to these comparisons, the automobile industry displays a number of distinctive characteristics, notwithstanding that it also shares a number of features with these other mature industries. These characteristics are as follows:

(1) The automobile is a consumer durable that meets some of the fundamental needs of the mass consumer. In this sense, it is distinctively different from steel or ships. Individual ownership of automobiles meets basic transportation needs, and the automobile has no serious competition for meeting the needs of the mass public for personal transportation. It is closely associated with the personal freedom of individuals to go where they will when they will. To be sure, mass transportation of one kind or another can fill important niches, but no form of mass transportation has yet evolved that seriously threatens the hegemony of the automobile in those societies where it has been established. As a product subject to the taste and aspirations of the mass public, we would expect that, in a competitive market, the automobile would change its shape and function according to the desire of consumers.

With these thoughts in mind, we suggest that the viability of the automobile industry *per se* is not in question, assuming its ability to service such basic needs. Yet the viability and growth rates of particular firms and national industries—and therefore the future pattern and structure of the industry—will be very much dependent on whether or not existing firms will be able to reasonably anticipate changes in consumer demands. In addition to "industry push," "consumer pull" is also important in determining future directions. Yet, the survival of particular firms and national industries requires integrating the results of past research and development with consumer demand. That is, technological innovation may be seen as antecedent to meeting consumer aspirations and central to determining the outcome of emerging competitive relationships among firms. To put the matter more sharply, such technological innovation can be a central feature in the renewal of the auto industry through the creation of new markets. Such a scenario is critical to reversing or at least slowing the decline in the growth of world automobile population that has occurred since the early 1960s (see chapter 6).

(2) The second characteristic observed in our research relates again to the consumer as a driving force in industry developments. Increasingly, in the industrialized nations, there is a diversification and volatility in consumer taste leading to rapid changes in consumer preferences, further market segmentation, and changes in production systems. In Japan, for example, there are increasing signs of a market segmentation between the inexpensive minicar market and the luxurious compact-car market.

We see the continued evolution of flexible manufacturing systems of subassemblies and finished vehicles—both *strategically* and in the specific sense of manufacturing technology—in order to better satisfy versatile and rapidly changing consumer tastes. This impacts the very organization of factories and the work process, manufacturer-supplier relations, and product-development strategies. For example, in order to reduce lead time for product development, greater stress is being placed on improving the coordination between product-development and manufacturing staffs, including the personnel of suppliers. This is particularly the case in the United States where longer product-development cycles have put producers at a competitive disadvantage. Computer-aided design (CAD) and computer-aided manufacturing (CAM) are important technological developments that have already shown their utility in significantly reducing lead times. In this area, it is the Japanese who are playing "catch-up." In still another important area, firms are reexamining their relative degree of vertical integration to meet the diversification of consumer taste and the need for greater flexibility. In Japan, we may see greater vertical integration as original equipment manufacturers (OEMs) move upstream in the production process in such areas as materials, particularly new materials. In the United States, the already highly integrated automobile firms seem to be moving toward less vertical integration in a search for greater economies and flexibility.

(3) There are many innovations occurring inside and along the periphery of the automobile industry, and there is no doubt that such changes can open up a new future in the fashion described above. One of the most revolutionary changes, for example, relates to the explosion of communications technology. This technology operates on the future of the automobile in two major respects. First, by altering the nature of work sites and the need to travel, it suggests the possibility of radical changes in one of the major functions of the automobile, commuting to work. The balance between transmitting people's thoughts versus people seems likely to shift. The notion of "electronic cottage industry" as a dominant form of work organization in the future is extreme, yet there is little doubt that communication technology has the potential to transform the geographical concentration of work organizations and, thereby, the role of the automobile. This is particularly the case in those countries such as the United States, where the combination of a geographically dispersed industry combined with a relatively weak public-transportation system often makes access to an automobile crucial to securing a job and, therefore, one's livelihood.

Second, the new communications technology means that the automobile is no longer a "closed box"; it is potentially able to communicate with the outside world anywhere at anytime. Telecommunication technology has made it possible, perhaps through satellite systems, for an automobile user to be interconnected with outside information networks. Due to the telecommunication revolution and the rapid progress in developing the social infrastructure appropriate to an advanced information system, the image of a car and its functions will be dramatically changed in the future.

Other revolutionary changes already occurring are in the field of automotive electronics, including the areas of engine and transmission control, diagnostics, etc. All these developments discussed above raise the further possibility that firms in the communications industry will play a major role in the evolution of the automobile industry, perhaps even to the detriment of existing automobile firms. Because those inside an industry tend to concentrate on marginal differences between their products and those of their competitors, major innovations often come from outside. While such interindustry competition is not uncommon, it usually comes as a complete surprise. For example, the chemical industry developed permanent-press washable fabrics and essentially eliminated the once thriving laundry business. One can also cite the impact of General Motors on existing steam-locomotive manufacturers or the impact of IBM on typewriter manufacturers.

Moreover, the quickening pace of technological innovation is not limited to electronics but spreads into other areas such as new materials and process technology. As William Abernathy and his associates note, the significance of technological change is the extent to which it disrupts established production, competence, marketing and distribution systems, capital equipment, organizational structures, and the skills of workers and managers. There is increasing evidence that this "disruption" is taking place in the auto industry, and the basis for competition is rapidly changing with it.

(4) The fourth characteristic notable in the automobile industry is the rapid internationalization of the industry in all its varied aspects. This interacts with the worldwide evolution of automotive and communications technology and the strategy of meeting diversified consumer tastes through worldwide production strategies. A variety of cooperative arrangements among the existing automobile companies, including the parts producers, ranging from merger, joint venture, and licensing in the design, production, and marketing phases are transforming the geography of automobile production. With the most rapidly growing markets expected to be in Third World countries, these countries will play a significant role in future developments.

Many of the developing countries have adopted restrictive practices designed to limit imports and thereby encourage local production. Indeed, there has been a growing sentiment worldwide that trans-national companies have an obligation to provide jobs, pay taxes, and support the economy of the developing nations they are serving. A number of the newly developing countries, such as Taiwan and Korea, are developing significant infrastructures for automobile production. The possibility of new entries into auto production in conjunction with cooperative tie-ups to existing producers appears quite likely over the next few decades. The role of new actors on the world automotive stage seems likely to range from those that are thoroughly integrated into the production strategy of existing producers (such as is currently the case with Brazil and Mexico) to those that will come to participate in a relatively more independent fashion.

In this era of growing integration of a worldwide industry, however, we are unlikely to see the entry of new, totally independent producers. As opposed to some of the mature "footloose" industries such as apparel, there is not likely to emerge a pattern of world-market penetration by a rapid succession of low-wage developing countries. The rising capital requirements, the increasingly demanding technology, and the need for large-scale management coordination of the design, production, and marketing of a product of considerable complexity operate to limit easy entry by low-wage, less-developed economies.

These four characteristics—the consumer base of the industry, flexible manufacturing strategies, rapidly evolving technology, and the internationalization of the industry—seem destined to transform the industry as we now know it. The existing automobile companies and their employees face a period of great uncertainty and transition. The competitive relationships among existing and new firms will be heightened at the same time that specific cooperative relationships develop. The stakes will be high for direct employees, indirect employees, consumers, shareholders, and the governments that benefit from the location of automotive companies within their national borders.

Throughout this study, we have seen already how these four driving forces are manifesting themselves in dramatic changes occurring in American and Japanese automobile firms. In addition, these dramatic changes are reflected in specific aspects of the respective industries. We note the following similarities and differences in the ongoing and expected changes in the two industries:

1. Technology

  United States a. Rapid introduction of production automation and reduction of in-process inventory.
          b. Major reductions of lead time for new product development (e.g., through the use of CAD/CAM).
          c. Appearance of new applications, products, and materials.
          d. Better integration of product and process engineering.

  Japan   a. Rapid introduction of production automation.
          b. Major reductions of lead time for new product development (e.g., through the use of CAD/CAM).
          c. Appearance of new applications, products, and materials.

2. Human Resources and Labor Relations

  United States a. Continuing attempt to upgrade sharply human resource training and utilization and expand union cooperation. This will proceed in a delicate balance with efforts to change work practices so that more efficient operations result.
          b. Growing recognition of the need to mobilize all human resources in competitive struggle, including blue-collar workers. Office automation will transform the number and deployment of white-collar employees. The organization of management will be restructured.

  Japan   a. New labor-management agreements regarding the introduction of robots and other labor-saving machinery.
          b. New strategies for coping with an aging labor force and the need to supervise operations and personnel in foreign countries.

3. Manufacturing

United States   a. Continued major restructuring of relationships, with reduction in numbers of existing suppliers accompanied by less vertical integration of OEMs. Closer long-term relationships with surviving suppliers.
  b. Increased sourcing from abroad.
  c. Continued strong emphasis on productivity and quality improvements.
  d. Gradual shrinkage of employment opportunities for blue-collar and white-collar employees with social and political consequences.

Japan   a. Rapid enhancement of technological capability of parts suppliers (including strong stress on the introduction of flexible manufacturing systems).
  b. Internationalization of the supplier industry.

4. Management Philosophy and External Environment

United States   a. Ongoing reconceptualization of the industry's competitive strategies, including the recentralization of domestic production operations in the Midwest and cooperative arrangements with other world producers in minimizing costs of bringing new products to market.

Japan   a. Diversification and differentiation of the domestic auto market (e.g., increasing segmentation between inexpensive minicar and luxurious compact car).
  b. Ongoing reconceptualization of the industry's competitive strategies, including the internationalization of production operations and cooperation with other producers, in response to trade conflict.

Within each category, the ongoing changes are not necessarily the same for every company. This is due to the different historical experiences and differing competitive niches of the two national industries and the different producers within each country. Generally speaking, changes in the United States reflect the need for producers to restructure their operations to more effectively compete both domestically and in worldwide markets. In Japan, the challenge is both to stay competitive and to cope with a more restrictive worldwide environment for Japanese imports. In the subsequent chapters, we will discuss in detail the various manifestations of this ongoing process.

Perspectives

The issue of the U.S. and Japanese automobile industries and their mutual relationships has to be examined in the context of the rapidly changing climate of the industry. Lack of information on what has been happening and on the nature of the transformation of the industry in recent years leads to misunderstanding and mistrust. Information sharing and frank discussion between both parties, Japan and the United States, is of great importance. This can be encouraged through the process of technical cooperation, marketing agreements, and other forms of cooperation across national boundaries. Yet, it is not to be thought that this is a simple problem that can be solved by increased communication. Quite naturally, interests can often be expected to differ, not only between the national industries but also among the various producers in both countries.

Automotive markets in the industrialized nations continue to shift from growth markets to primarily replacement markets. Furthermore, they have become more transitional in nature. Under such circumstances, the competitive outcome—and the direction of trade imbalances—will be very much conditioned by the capacity of the automobile companies in the two countries to produce competitive cars to meet market demands. This, in turn, will depend to a significant extent on the technological capability of given automobile companies. To be sure, wage rates, exchange rates, production location, and other factors impact on the ability of companies to produce competitive cars. Yet, the technological factor, broadly conceived, is a critical one in our judgment.

Judging from past performance and statistics of research and development (R&D) expenditure and expenditures on research manpower, there is no doubt that the technological level of U.S. and Japanese automobile companies is high. The capacity for technological innovation of both industries stands on a world-class level.

In the U.S. market, the recent, modest shift of U.S. consumers to larger cars—in the context of reduced gasoline prices—suggests that the U.S. automobile industry will maintain its competitiveness in this traditional, though quite restricted, market. However, with regard to the world market, and presumably also in the United States, it is to be expected that the strong demand for small-sized cars will continue in the future. Therefore, the real competition in the auto market will be in this category.

Increasingly, competition may be between enterprises and coalitions of enterprises rather than national-flag industries. These cooperative relationships and coalitions will evolve as a strategy to exploit new technologies and to strengthen competitiveness in the changing environment of the automobile market. Expressed differently, they are strategies to reduce uncertainty in a rapidly changing environment.

In essence, this movement toward strengthened corporate linkages derives primarily from two factors. First, no automobile company can monopolize a superior position in the world automotive market. This is the case because rapidly changing technology in the hands of competitors constantly threatens established positions. New materials, new manufacturing technology, the possibility of shifting energy sources, and the expanded role of electronics are but specific manifestations of this dynamic state.

Second, during this transitional stage a large R&D investment is needed. Yet, with the volatility and diversification of consumer preference, the life of specific products is shortened. Under these conditions, the automobile companies must find a way to compensate for the huge investments that are required while still maintaining sufficient variety to meet varied consumer demands. The obvious solution is to expand sales of specific models, and cross-national cooperation among producers is seen as an important strategy to assure a larger market for products.

History suggests that, in the long run, one must be concerned that this pattern does not lead to a cartelization of world markets that would end up diminishing competition. For the time being, however, on a worldwide scale, vigorous competition among the emerging coalitions appears the norm. Restriction of competition, whether from cartelization or import restrictions, would lead to a decline in technological innovation and, therefore, in the incentive of manufacturers to service consumer demands in a timely fashion.

Yet, one cannot ignore the existence of crucial macro political and economic problems, such as national and international economic slumps, rapidly shifting trade imbalances as exchange rates respond to surges of financial flows, and the process of politicization of the issues that inevitably accompany such events as rising unemployment levels. While various measures to accommodate to these pressures will understandably be taken, these measures must not be allowed to protect permanently the industries of the advanced industrial nations. To allow this to happen not only would weaken the technological competitiveness of the automotive industry in the specific country in question but also would discourage the ongoing dramatic changes in management philosophies, human resource development and labor relations, manufacturer-supplier relations, and manufacturing processes. Such permanent protection would lead to a choking of the future potential of the industry and cause it to decline along the lines of the steel and shipbuilding industries.

A related outcome of such stagnation in technological progress would be that the comparative advantage in automobile production currently held by the industrialized nations would quickly dissipate. We cannot afford to indulge ourselves in the mistaken "Suez mentality" that assumes that the less-developed nations do not have the capacity to learn to operate complicated technological processes. Thus, technological progress in the

automobile industry of the industrialized nations must go forward lest the
industry be caught up in a vicious cycle of decline. Such a decline would invite
still greater policy interventions on behalf of protectionism, thereby further
weakening the competitiveness of existing firms. The core of the dilemma
facing industrialized nations is how to make the adjustments to the macro
political and economic problems without succumbing to a vicious cycle
inviting restrictions, stagnation, and decline.

Chapter 2
Postwar Evolution of the Japanese and American Automotive Industries

This chapter is intended to provide an introduction to the specific aspects of the automobile industry in Japan and the United States, as well as general background information. We will deal with the impact of the auto industry and its role in the respective national economies. This will involve treating the transformation of the respective auto markets, including overseas sales. We will also examine the degree to which auto firms are embedded in larger financial and manufacturing institutions. With regard to the environment in which auto firms operate, we will look at the evolution of government policy making as it has impacted on the auto industry. A major focus will be the evolving characteristics of the respective industries. Among the internal characteristics to be considered will be the changing organizational structure of auto manufacturers, the development of manufacturer-supplier relations, and technological changes in the postwar period. For the most part, our treatment will focus on the postwar period, but, where necessary, earlier periods will be discussed to provide the reader with a broader perspective.[1]

The U.S. and Japanese auto industries account for some 50 percent of total world auto production, and their domestic markets account for 45 percent of all passenger cars registered in the world. Characteristics associated with their rise to prominence are, therefore, of major interest.

Automotive Impact

Since the start of the mass production of automobiles, American life has become increasingly dependent on and shaped by motor vehicles. Motorized transportation has made the United States the most mobile society in the world. Tractors and mechanized equipment have revolutionized farming, and motor vehicles have dramatically affected the design of homes, cities, and

---

[1]The following sources were particularly useful in providing background data: Japan Automotive Manufacturers Association, *Jidōsha Tōkei Nenpō* [Yearbook of automotive industry statistics] (Tokyo: Japan Automobile Manufacturers Association, annual issues); Motor Vehicle Manufacturers Association, *Facts and Figures '82: Strategy for Recovery* (Detroit: Motor Vehicle Manufacturers Association, 1982) and *World Motor Vehicle Data* (Detroit: Motor Vehicle Manufacturers Association, annual issues).

rural areas. They have become a pervasive and generally positive factor of daily life for nearly every person, group, business, and government in the nation.

Most U.S. households have a car or a truck. In 1977, less than 14 percent of all households in the United States did not have a vehicle. Over 14 percent of households own 3 or more vehicles. On an aggregate basis, the United States has 537 passenger cars and 157 commercial vehicles—for a total of 694 vehicles—for each thousand of population.

In contrast to the U.S. situation, the Japanese, until recently, perceived a passenger car as a luxurious commodity, far beyond the reach of the general public. This perception stems from the mid-Meiji era when only the aristocrats enjoyed a highly westernized, extravagant motor culture. In fact, except for the short period of the "minimotorization" in the 1930s, the Japanese thought of motor vehicles as military and civilian trucks and buses, or taxicabs. Before the late 1950s, it was inconceivable to think that almost everybody could afford to own a car.

Primarily due to the sucess of the famous "Income Doubling Plan" of the Ikeda Cabinet, the price of passenger cars fell within reach of the middle class in the late 1960s, at which time mass motorization began. At present, as of the end of fiscal year 1982, only 31.7 percent of all households in Japan were without a passenger car. Per capita income in this same year rose to $7,100.[2]

In Japan there are 209 passenger cars and 126 commercial vehicles per thousand population, or a total of 335. Compared to the United States, Japan has approximately half (48 percent) as many vehicles per thousand population, but this is made up of an 80 percent level for commercial vehicles and only 39 percent at the passenger-car level. Western Europe shows a similar total, with 49 percent as many vehicles per population as in the United States, but the pattern there is 22 percent at the commercial level and 57 percent as many passenger cars. Canada, which is perhaps most like the United States in geographic use and need of vehicles, has identical ratios of 81 percent as many passenger cars, commercial vehicles, and total vehicles as in the United States. Similarly, the ratios worldwide are even, although at a low level of only 14 percent as many vehicles per thousand population as in the United States. Canada and Japan have the same number of commercial vehicles per thousand population.

In the United States, private motor vehicles are by far the main means of transportation to work. In 1980 over 86 percent of U.S. workers relied on a private motor vehicle to get to work. Another 7.4 percent walked or used other means, while only 6.4 percent used public transportation. (The public transportation fraction has been declining over the years.)

---

[2]Calculated at the rate of ¥240 = $1. This conversion rate will be used throughout the report unless otherwise indicated.

The contrast in the role of the motor vehicle as a means of commuting to work in Japan is quite striking. According to Japanese 1970 census data, only 14.5 percent of employed commuting persons and students over fifteen years old used a private car to get to work. Twenty-three percent walked; 29 percent used the railway, subway, or street car; 14.6 percent used buses; 1 percent taxis; 2.4 percent a company or school bus; and 15 percent a bicycle or motorcycle. At that time, an additional 6 percent used autos for business purposes. By 1979, the privately owned auto had become a more popular mode of transportation to work, with some 22 percent of those working and attending school over age fifteen now relying on privately owned cars, and with another 7.4 percent using autos for business purposes. Yet this is still far below the 86 percent of American workers who rely on private cars to commute to work. Although the utility of cars for the Japanese cannot be underestimated, cars do not serve the same function as in the United States. These kinds of differences reflect primarily the geographical dispersion of population and, therefore, the feasibility of large-scale public transportation, government policy as it impacts on car ownership, and per capita income.

Long-Term Trends in Automobile Production

American production of motor vehicles, including passenger cars, trucks, and buses, was nearly 8 million units in 1950 and reached a peak of almost 13 million units in 1977. Increasing international competition and the long worldwide economic downturn of the early 1980s led to a decline to a little over 7 million units in 1982. This pattern is shown in figure 1. It also indicates the large year-to-year fluctuations of production in the United States.

Also shown is the growth of Japanese motor-vehicle production beginning in 1950. Production first passed the 1 million mark in 1963, having increased sharply from the very low levels of the early 1950s. It rose almost steadily to a peak of over 11 million units in 1981 and then declined to 10.7 million units in 1982. Unlike the United States, we see a fairly steady growth pattern; this has important implications for planning and resource allocation.

The smaller volumes of the late 1940s and early 1950s understate the role of automobiles, or at least four-wheeled vehicles, in prewar Japan. General Motors and Ford played a dominant role in supplying cars to Japan in the prewar period, but legislation enacted in 1936 effectively removed them as factors. By 1939 Nissan and Toyota were producing almost 30,000 vehicles per year, but they were mostly commercial vehicles, not passenger cars.

American production accounted for 76 percent of world production in 1950, declining rapidly to 45.4 percent in 1965, with the bulk of the decline accounted for primarily by the rise of Western European producers. By 1981, the United States accounted for 21.2 percent of world production. Japan accounted for 1 percent of world production in 1956, 7.6 percent in 1965, 20.9

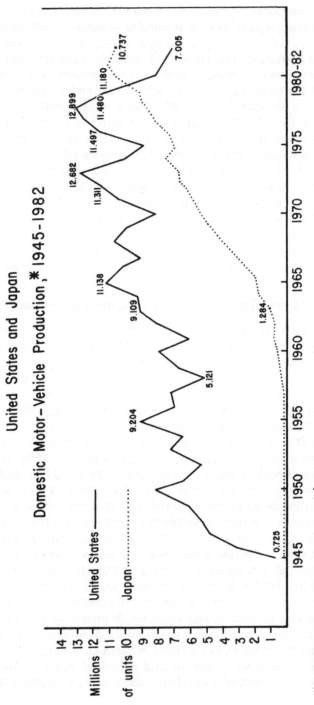

FIGURE I

United States and Japan

Domestic Motor-Vehicle Production *, 1945-1982

* Includes passenger cars, trucks, and buses.

percent in 1975, and 30 percent in 1981, a remarkable rise to world prominence in such a short time.

It is instructive to compare the first nine active years in the histories of both the U.S. and Japanese automotive industries after they achieved annual production in excess of 100,000 units (see figure 2). The U.S. industry first reached this level in 1909, and nine "normal" years later (excluding the 1918 war year) it achieved production of 1,876,000 units. The Japanese automotive industry first produced in excess of 100,000 units in 1956, and nine years later it reached the identical annual level of 1,876,000. While this suggests similar early growth rates, we should also note that the U.S. industry, starting in 1909, had twenty-five new highs over a seventy-year span, culminating in the almost 13 million units produced in 1978. In only twenty-six years, between 1956 and 1981, the Japanese industry had twenty-four new highs and reached the level of a little over 11 million units.

Trends in Auto Markets

The total number of automobiles (or automobiles plus light trucks) is a stable function of key economic variables such as real income, driving-age population, the quality-adjusted price of cars, and the price of gasoline. Auto ownership in the United States has increased steadily over the last thirty years, but at a declining rate. The total stock of cars increased at 4.3 percent in the 1950s, at 3.5 percent in the 1960s, and only 2.6 percent in the 1970s. Growth in the total stock of cars in the United States will be even lower in the 1980s, probably about 2 percent per year. With historical scrappage rates, this implies somewhat more than 11 million sales per year in the 1980s. One important consequence of this declining rate of increase in the demand for the total number of automobiles (which is much more stable and predictable than the demand for particular subcategories such as small cars or new cars) is that new cars will be used more than ever before to replace old cars. Consequently, the relative characteristics of old and new cars will play a more important role in the auto market.

The Japanese auto market has displayed considerably greater growth than the U.S. market. For example, between 1975 and 1980 total car registrations grew at an annual rate of 6.8 percent. This is likely to continue since the United States has 55 cars for every 100 inhabitants, while Japan has only 20 cars per 100 inhabitants. Consequently, the Japanese market, although characterized by increasing replacement demand, is more nearly like the U.S. market of the 1950s. This is tempered somewhat by a number of factors. The higher population density of Japan suggests a lower saturation level (see chapter 6 for an evaluation of this as well as an alternative view). There is also a lower life expectancy for Japanese vehicles. Currently, median life expectancy in the United States is over ten years, compared to just under

FIGURE 2

U.S. and Japan: Record Years in Motor-Vehicle Production[a]

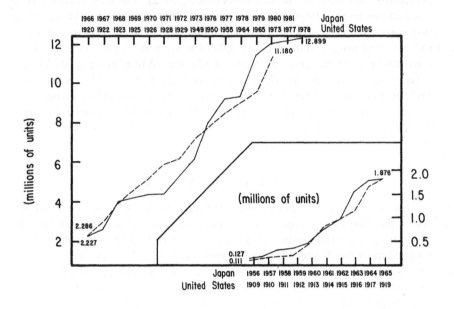

—————— UNITED STATES

[a] Cars, trucks, and buses.

[b] First-year production exceeded 100,000 units.

– – – – – – – – – JAPAN

	U.S.	JAPAN
Number of new highs:	25	24
Beginning year[b]:	1909	1956
Ending year:	1978	1981
Span of years:	70	26

eight years for Japan. Moreover, the high growth rates of the total auto stock in Japan are expected to decline markedly, with total vehicle registrations possibly leveling off at 50 million. The total number of vehicles in use in Japan in 1982 was 41 million. This expectation is natural in light of the higher population density and far better mass-transportation system in Japan, but it may be somewhat elastic in view of the possibility of using smaller vehicles in nontraditional ways, a trend currently under way in Japan (see chapter 6).

Many analysts currently expect a gradual decrease in the share of the American market for full-size and intermediate passenger cars from the current 50 percent to probably about 25 percent of the market by 1990, and a corresponding increase of compact and subcompact (and smaller-size) cars. Similar trends in Japan are also expected and are in fact already under way. Sales in light-weight trucks are already 25 percent of the market, and the introduction of new light-weight models is expected to encourage this trend. Undoubtedly, rising energy prices and, in Japan, an increase in the number of two- and three-car families have stimulated the move to smaller vehicles. One issue of key importance, the market share of imports in the United States, is difficult to predict. Studies made in 1979 and 1981 forecast total foreign sales (imports and domestic production of foreign firms) at 24 percent or less through 1990. Import sales in the early 1980s have been well above this figure, however, suggesting the difficulty of arriving at accurate predictions.

Automobile Industry Concentration

Some 181 different companies actually manufactured passenger cars commercially in the United States during the period 1903 through 1926. Of these, 137 had either gone out of business or had been acquired by a competitor during the period, and 44 remained in 1926. The Census of Manufacturers listed 40 separate producers of passenger vehicles in 1927. The 1935 census reported 17 manufacturers (not including makers of buses and taxicabs), and the Automobile Manufacturers Association reported only 11 companies manufacturing gasoline-powered passenger cars in 1939. At the beginning of the postwar period, there were 10 companies manufacturing passenger cars (see lower right segment of figure 3). The total declined to 5 by the late 1950s and, including Volkswagen of American, stood at 5 in 1980. The addition of Honda in 1983 increases the total to 6.

Paralleling this fairly rapid decrease in companies producing cars, the market share achieved by the Big Three (General Motors, Ford, Chrysler) is reported in the upper segment of figure 3. It increased from 36 percent in 1910 to 87 percent in the 1930s, rising eventually to 98 percent. (These shares refer only to the production and sales of domestic vehicles.) There has been a similar shrinkage in the size of the U.S. retail distribution system. From approximately 30,000 new domestic car dealers in the prewar period, the

FIGURE 3

Big Three[1] Share of Domestic Market[2] and the Number[3] of Companies Manufacturing Passenger Cars in the U.S.

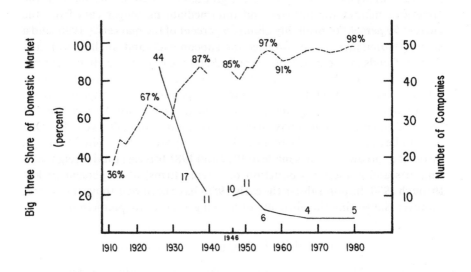

--------- BIG THREE SHARE
                OF DOMESTIC MARKET

——— NUMBER OF COMPANIES
              MANUFACTURING
              PASSENGER CARS
              IN THE U.S.

[1]General Motors, Ford, and Chrysler.

[2]Production, 1911-21; sales, 1923-81.

[3]Excludes Checker and smaller companies that have, in total, accounted for less than 1 percent of registrations since 1946 and 0.1 percent or less since 1963.

number of domestic dealers rose to about 50,000 in the early postwar period, but declined to 21,000 in early 1984. During this period of decline, the annual sales per outlet rose from levels near 100 cars per dealer to over 400 in the peak years of the late 1970s. In addition, Japanese manufacturers now have almost 5,000 dealers in the United States. This excludes domestic dealers selling only those Japanese products, such as the Dodge Colt, distributed by domestic manufacturers. Over one-third of the Japanese dealers are dualed with domestic makes.

With regard to Japan, the Law of Support for Military Vehicles of 1926 acted to reorganize the Japanese auto industry, which was previously composed of small-size manufacturers. As a result, three major companies, Tokyo Gas & Electric, Ishikawajima, and DAT Motors, emerged. In the 1930s, there were six primary manufacturers: Nissan, Mitsubishi, Japan Ford, Japan GM, Automobile Industrial, and a new entry, Toyota Motors. Fear of domination by the two powerful foreign entries, Japan Ford and Japan GM, led to the promulgation of the Law Regarding Automobile Manufacturing Enterprise of 1936. This had the effect of eliminating Ford and GM from the market.

In the postwar era, the Ministry of International Trade and Industry (MITI) tried, with only modest success, to attain the optimal number of auto manufacturers by merger under the "rationalization" requirement and in view of the forthcoming trade liberalization of the mid and late 1960s. The failure of MITI's People's Car Plan indirectly aided the new entries of minicar producers, adding to the existing two powerful manufacturers, Toyota and Nissan. Putting aside the argument on the optimal size of the auto industry under economic rationalization, it is clear that the new entries of the relatively weak manufacturers intensified competition among the existing manufacturers in technology and product development. Currently, there are nine independent Japanese automobile manufacturers.

The Japanese retail distribution system was originally modeled on that of General Motors and Ford, based on their pre-World War II distribution network in Japan.[3] This system was gradually modified to fit Japanese needs. In particular, close tie-ups between manufacturers and the dealers were introduced in the form of affiliations. These relationships were reinforced by the establishment of a manufacturer-led sales-financing system set up to compensate for the underdevelopment of consumer-financing institutions. Manufacturers took the lead in helping dealers establish modern sales techniques. The number of dealer outlets rose rapidly in postwar Japan, increasing by the mid-1970s to 1,430 main franchises with 10,153 sales locations and a total of 305,000 employees. By 1983, the number of main

---

[3]Akoi Okochi and K. Shimokawa, *Development of Mass Marketing: The Automobile and Retailing Industries* (Tokyo: University of Tokyo Press, 1981).

franchises rose to 1,768 with 14,185 sales locations and 335,000 employees.[4] Sales territories have commonly been large, with a high number of units being sold per dealer. In 1971, the main franchise dealers sold an average of 1,499 vehicles per dealer; this rose to 1,704 in 1982. For the major producers, this average is much higher (4,300 for Toyota dealers in 1978).

Economic Impact of the Automobile Industry

The role of the automobile industry in the respective economies is a major consideration. In 1982 an estimated 3.72 million people owed their employment directly to the U.S. auto industry. When all automotive-related and dependent activities are considered, the number was on the order of one in ten jobs in the country. Prior to the sharp downturn in production in late 1979, the auto industry typically consumed 21 percent of the nation's steel production, 60 percent of its synthetic rubber, 11 percent of its aluminum, 30 percent of its iron castings, 25 percent of its glass output, and 20 percent of machine tools. There are over 2,000 supplier companies primarily dependent on the automotive business for their success and 40,000 in total providing goods and services. In the late 1970s, vehicle manufacturers were purchasing materials, components, and services in excess of $55 billion per year. All indications are that the total number of suppliers is shrinking in view of the protracted economic downturn, announced policies on the part of the vehicle manufacturers to shrink their supplier base, and attempts by many supplier firms to turn to more attractive growth areas. Approximately 30 percent of all retail sales are automotive vehicles or related parts and service. This includes, of course, sales of used vehicles, service, and "aftermarket" sales of parts. Finally, the industry is immeasurably important to national security.[5]

If we examine the breakdown of direct employment in vehicle manufacturing and related industry, we find that in 1982 vehicle manufacturers (car and light truck) employed 685,000 persons, and suppliers added an additional 935,000. In addition, there were almost 2.1 million jobs in sales and service. This total of some 3.72 million jobs represented about 3.9 percent of total employment in 1982, and 5.0 percent of all employed wage and salary workers in the nonagricultural sector. In the manufacturing sector, automotive employees accounted for 19.7 percent of all employees. Moreover, the figures just cited do not include those employed in the fuel-supply and transportation sectors (e.g., truck drivers and employees of auto-rental firms). Were these more indirectly related employees included, the industry would account for

---

[4]The totals for sales locations only, including sites for used cars.

[5]For further discussion of data in this section, see *The U.S. Automobile Industry, 1980* and *The U.S. Automobile Industry, 1981,* Reports to the President from the Secretary of Transportation (Washington, DC: U.S. Department of Transportation, Transportation Systems Center, 1981 and 1982, respectively).

one in ten employees in the total economy. The protracted economic contraction of the industry from 1979-82 led to sharp reductions in vehicle manufacturer and supplier employment with an estimated 27 percent loss of jobs. Whereas the automotive industry was consuming 21 percent of all domestic steel production in the past, shipments to the auto industry fell to 15 percent by 1982. Although an extended rise in sales and production would restore many jobs, it is widely agreed that many of those laid off will never return to the automotive industry.

The economic impact of the industry can also be seen in the annual investment in new plant and equipment expenditures by vehicle manufacturers and their suppliers. Despite cyclical swings, the increasing size of investment required to keep the domestic auto industry competitive is notable. These increases are consistent with those of other auto firms throughout the world. Moreover, despite depressed sales and production in 1980-82, the manufacturers and suppliers invested in new capital at a rate of $9 billion (nominal) per year. Expressed in 1982 dollars, the auto industry annually averaged $7.1 billion for new plant and equipment expenditures from 1970-76. From 1977-82, the average rose to $9.9 billion. In the recession years of 1981-82, the auto firms accounted for an average of 2.8 percent of the total investment in expansion and modernization projects by American business and 7.3 percent of all manufacturing investment. Yet, the scale of the necessary investments combined with depressed earnings contributed to a substantial weakening of the financial strength of the U.S. auto companies.

Workers in automotive and automotive-related industries in Japan also account for about 10 percent of Japan's total labor force. A little over 7 percent are in directly related fields (according to MITI's classification). This includes 696,000 in automobile manufacturing (OEMs, parts suppliers, body industry, and tire industry), 248,000 in the fuel-supply field (gasoline stations), 997,500 in the sales and maintenance area (auto dealers, parts and accessory wholesalers, retailers, and service shops), and 1,594,000 in the utilization area (passenger and freight transport, auto rental). Those indirectly employed in the industry include 713,000 in materials supply (paint, glass, steel, and nonferrous metal manufacturers), 127,000 in fuel supply (petroleum refiners, lubricants and grease manufacturers, and petroleum wholesalers), and 515,000 in supporting industries (insurance, advertising, and other transportation-related services). According to 1982 statistics, the Japanese automotive industry yielded $99 billion; this total accounted for 30 percent of the value of Japan's total production of the machinery industries. The industry's share of total exports is 20 percent, totaling $30.2 billion. This amount is twice as much as accounted for by steel exports and 60 percent of Japan's total imports of crude oil.

When we examine capital investment in plants and equipment, we find the role of the automotive industry in Japan's total economy also to be of major, if not even greater, importance. Japanese auto-industry capital investment is running at a far lower level than in the United States, reflecting both the greater need of the Americans to reorient their product offerings and perhaps a more efficient use of capital on the part of the Japanese. Yet, expressed in 1982 dollars (at an exchange rate of ¥249.05 = $1), investment in Japanese auto plants and equipment tripled from the mid-1970s to an average of $3.4 billion for 1981 and 1982.[6] The Japanese auto industry accounts for some 20 percent of new plant and equipment investment in the manufacturing sector. This represents almost a doubling from 1976, when it accounted for 12.7 percent of manufacturing investment, and it also is a far higher percentage of manufacturing investment than is accounted for by U.S. auto firms. There is no doubt that the Japanese automotive industry today contributes significantly to the national economy.

Evolving Strategy for Overseas Sales

In 1981, Japan accounted for 31.9 percent of world trade in motor vehicles, with the United States accounting for 7.2 percent. An alternative picture of these trade flows may be gained by eliminating intra-European and U.S.-Canadian trade where special rules facilitate trade between the United States and Canada and among European Community (EC) members. Under this definition, the Japanese share is 70.7 percent, and the U.S. share is 2.5 percent (see chapter 3 for further detail). Over 40 percent of Japan's passenger-car exports go to North America, but a miniscule number of U.S. exports go to Japan (under 4,000 vehicles in 1982). Japan exports some 50 percent of its passenger-car production (comparable to many Western European countries), and imports have never exceeded 2 percent of total new-car registrations in a given year. Imports accounted for 27.9 percent of the U.S. market in 1982, with the Japanese accounting for 81.1 percent of total imports.

These data demonstrate radically different strategies for capturing world markets by Japanese and U.S. manufacturers. American manufacturers have historically exported capital in the form of creating foreign production facilities rather than exporting finished vehicles. Ford Motor Company established this practice with its construction of the first overseas assembly plant in England in 1911. Incentives included lower shipping costs and lower import duties. This pattern of development continued worldwide. In 1982, Ford sold 2 million cars and trucks outside North America, compared to 2.3 in North America. In this same year, Ford achieved a higher market share in

---

[6]These figures and the national comparisons based on them should be taken to provide only the most elementary description since the Japanese data are limited to thirteen manufacturers (eleven automotive and two motorcycle) and do not include the investments in plant and equipment of material and parts suppliers.

the United Kingdom, Brazil, Argentina, and Australia than it did in the United States. For the eleventh consecutive year it led all manufacturers worldwide in vehicle sales outside their domestic production base. General Motors has followed the Ford pattern, though on a lesser scale, selling some 1.7 million cars and trucks outside the United States and Canada in 1982.

At first, the cars produced overseas by U.S. manufacturers were the same as their domestic models. Gradually, modifications were introduced until today overseas products are designed specifically for their local markets. The Ford and General Motors strategy has been one of producing locally to meet local needs.

In contrast to the practice by Western multi-national corporations of establishing overseas plants and operations, the Japanese automobile industry, with some exceptions, did not have a well-designed strategy for overseas operations. Rather, a focus developed on exports. In the early postwar period, the Japanese production and export capacity was limited primarily to trucks, and its exports were geographically limited to Asian countries. As figure 4 vividly indicates, the dynamic structure of the export curve of passenger cars has had three conspicuous trend changes, occurring in 1972, 1975, and 1979. The 1972 leveling off of passenger-car exports may have reflected the general economic slump worldwide. The 1975 and 1979 upward movements in exports were due to the oil crises and the rise in demand for small, fuel-efficient cars. In comparison with the dynamic growth pattern of the production curve, this export curve indicates that the boom of the Japanese fuel-efficient subcompact cars was initially triggered by an external causal factor, i.e., the oil crises. It would appear that, primarily because of the oil shocks, the characteristics of the Japanese passenger cars came to be well appreciated abroad, and, concurrently, a well-designed overseas sales strategy developed.

Embeddedness of Auto Firms in Larger Financial and Manufacturing Institutions

In the United States, there are a number of relatively small producers of motor vehicles (primarily trucks) that are part of larger organizations. The Big Three historically have been independent organizations. In its early years, General Motors had close ties with Dupont, and, indeed, during the 1921 crisis, Dupont stepped in to help reorganize the company in order to protect its investment.

Chrysler could not be thought of as an entirely independent company these last few years in view of the unusual and extensive loan program incorporating government guarantees and certain operating restrictions. But this support was intended to be temporary, and Chrysler has now repaid its government loans in full.

FIGURE 4

Production and Export of Japanese Passenger Cars

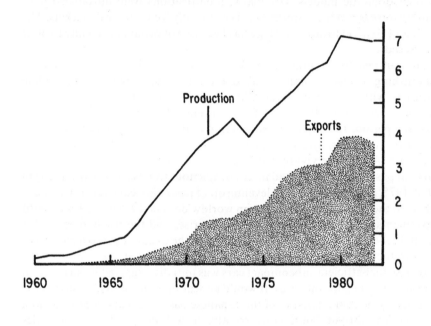

YEAR

This general pattern of independence from other manufacturing firms sets the U.S. industry somewhat apart from many Western European or Japanese counterparts, where such relationships with either the government (Europe) or other firms (Europe and Japan) are more commonplace. In general, Chysler, Ford, and General Motors borrow from banks and through other means in the normal course of business, but, with the exception of the presumably one-time Chrysler arrangement described above, there is no management participation by those making the loans.

Over 90 percent of the Big Three production and sales is automotive, although actual volumes of nonautomotive activities are impressive—over $3.3 billion for GM and almost $3.4 billion for Ford in 1982. Although the Big Three have a number of joint arrangements with Japanese and other manufacturers, these are not as yet large factors in their overall operations.

While such independence and concentration of activities is clearly a source of strength, it also means that in times of difficulty in the auto business the U.S. firms cannot as easily rely on other sources of business or support from "parent" organizations. American Motors Corporation represents an interesting situation where Renault has taken over effective control of the company. Thus, they have been able to preserve their manufacturing operations in the United States. Volkswagen of America is owned 100 percent by the parent company.

With the democratization of the Japanese economy after World War II, the *zaibatsu* industrial groupings that had played a major role in guiding the Japanese prewar economy were dissolved. This was in keeping with the newly enacted Anti-Monopoly Law. As the economy recovered, however, new industrial groupings (often based in part on prewar aggregations) began to form. The high cost of obtaining scarce capital and, still later, capital to fund rapid growth were major incentives. Banks and trading companies came to form core elements in these developments. Typically, the groups also included an insurance company, real estate company, and a number of manufacturing firms specializing in different product areas. Extensive cross-holding of stock shares cements these relationships, with the leaders of the member companies commonly meeting on a regular basis to discuss and formulate policies.

The auto companies, in addition to forming their own industrial groupings of suppliers and subsidiaries, have participated in these developments in varying degrees. Mitsubishi Motors is 85 percent owned by Mitsubishi Heavy Industries, while Tōyō Kōgyō is closely associated with the Sumitomo group. These groups display considerable cohesiveness, and this adds an important element of competitive strength to their members. Toyota has friendly relations with the Mitsui Group, one of Japan's two largest financial conglomerates. Toyota's position as the No. 1 company in Japan, however, means that it has a sound financial basis; its equity ratio is exceptionally high. Nissan is closely associated with the Industrial Bank, which carried out a major rescue mission of Nissan when it was on the verge of bankruptcy in the 1950s. Indeed, Chairman Katsuji Kawamata, who guided the company to its remarkable achievements in the next two decades, was a former official of the Industrial Bank. Some companies have looser ties—for example, Honda being known as "Mitsubishi colored"; Mitsubishi Bank is Honda's largest shareholder, holding 4.6 percent of its stock.

There are many important functions of such groupings. The provision of capital is less important today than in the early capital-short postwar years, but the large capital requirements of auto manufacturers in this new era of internationalization still make it critical for most auto manufacturers. Companies are more likely to cooperate with other members of their groups in developing new products and technologies. Electronics and process technology in the auto sector would be prime areas.[7] Mitsubishi Motors has been prominent in developing turbochargers based on the expertise of other Mitsubishi companies in the aircraft industry. (This and its truck operations were also the base of Saab's prominence in this area.) More generally, membership in these groupings provides important resources with regard to an information pool on technology, organization of management, sales strategies, overseas ventures, etc., which is an important asset in formulating corporate competitive strategies. In cases of financial difficulties, these groups can provide support that serves as a collective assistance effort with shared risks. The rescue of Tōyō Kōgyō by the Sumitomo group represents a case in point. None of this is to suggest that the auto firms are commonly subordinate to these larger groupings; rather, they should be seen as important support institutions that have added to the competitive edge of Japanese auto companies. It should be noted also that such groupings are not unique in the sense that a number of European auto companies (e.g., Saab and Volvo) are members of larger industrial groupings as well.

In examining the degree of independence of various operating units in automotive manufacturing, it is important to consider the supplier industry as well. American motor-vehicle manufacturers are highly integrated and include within their corporate structures many divisions and other wholly owned operations that supply parts and components for assembly operations. By comparison, many of these divisions might be called first-tier suppliers in Japan and stand formally as independent companies. Commonly, however, the Japanese vehicle manufacturers will possess some equity in these first-tier suppliers, and they may be no less integrated (in some cases perhaps more so) into the vehicle manufacturer's operations.

---

[7]James McArdle and Associates, *The Japanese Automobile Challenge* (London: James McArdle and Associates, 1982).

Development of Internal Organizational Characteristics

Both Ford Motor Company and General Motors have undergone momentous organizational changes in the course of their growth. Indeed, they both served in a variety of respects as models for other industrial organizations and, in that sense, shaped the history of manufacturing firms in the United States.

For an older generation, Ford was literally synonymous with the term automobile. With the dominant and lasting role of individual family members in starting and sustaining the company, many aspects of its organizational structure had more in common with 19th-century firms than those of the present century. Perhaps in response to limited capital, the early Ford production facilities were limited, with the capital costs and high overheads of an integrated factory being avoided through subcontracting. The small labor force and plant was focused on assembly operations. In this regard, it paralleled the strategy of the contemporary Japanese auto companies. With the success of the Model T, Ford transformed the auto business from a small-scale industry into a mass-market industry. Its strategy was a standardized model, efficient large-scale output, with a reduced retail price and higher and higher volumes of production.

As Ford grew, it pioneered in a variety of areas. Notable is the development of the branch assembly plant in the major market areas of the country and eventually the world market. There were substantial savings in freight costs by shipping parts instead of assembled vehicles. During the early 1920s, the highly vertically integrated manufacturing facilities that grew up around Dearborn became a model of material handling and factory organization. In the late 1920s and 1930s, Ford experienced a variety of crises that led to a decline in its market position. The personal leadership of Henry Ford worked fairly well when corporate structures and objectives were relatively simple. But as the firm grew and expanded into a variety of areas, the absence of a systematic management organization became an increasing liability. It was not until the early postwar years that a new management structure was adopted, partly based on the concepts developed at General Motors. New managers were brought in, and manufacturing activities became more decentralized (while assembly operations gradually became more concentrated).[8]

The organizational innovations developed at General Motors during the course of its history had a lasting impact not only on other auto firms but on U.S. business organizations in general. After the founding of General Motors in 1908, William Durant carried out a strategy of growth based primarily on

---

[8]This section draws extensively upon Gerald Bloomfield, *The World Automotive Industry* (London: David and Charles, 1978).

acquisition. He made relatively little effort to bring these many activities under centralized control.

Of particular importance for the evolution of the firm was the response to the post-World War I recession. In response to an inventory crisis, Pierre Dupont entered the company and worked closely with Alfred Sloan to rehabilitate the firm. They created what has come to be known as the multi-divisional structure. Alfred Chandler, Jr., the noted student of this process, describes its functioning as follows:

> Autonomous divisions continued to integrate production and distribution by coordinating flows from suppliers to consumers in different, clearly defined markets. The divisions, headed by middle managers, administered their functional activities through departments. . . . A general office of top managers, assisted by large financial and administrative staffs, supervised these multifunctional divisions. The general office monitored the divisions to be sure that their flows were tuned to fluctuations in demand and that they had comparable policies in personnel, research, purchasing, and other functional activities. The top managers evaluated the financial and market performance of the divisions. Most important of all, they concentrated on planning and allocating resources.[9]

This mode of organization seemed particularly applicable to auto firms manufacturing several lines for a variety of products. One can point out that over the years some backsliding has occurred (as, for example, when top management becomes involved with day-to-day operating responsibilities) and excessive staffing of the corporate office has created rigidities and a stifling of divisional initiatives. Yet there can be little doubt that the structure has served General Motors and other companies quite well during the course of their history.

Both Ford and General Motors can be characterized largely as decentralized companies. Each employs a profit-center concept and delegates decision-making authority to line management, although some variations exist as to the form of decentralization practiced. For example, General Motors's vehicle divisions (e.g., Chevrolet, Oldsmobile, etc.) operate with greater autonomy than divisional activities within Ford's North American Automotive Operations. General Motors's divisions have broader functional responsibilities and, therefore, can operate more as independent business units. On the other hand, Ford's Diversified Products Operations (DPO) divisions have substantial autonomy, function as individual businesses, and operate on a worldwide basis. Both companies provide some services on a centralized basis (e.g., certain purchasing functions, air transportation), with

---

[9]Alfred Chandler, Jr., *The Visible Hand: The Managerial Revolution in American Business* (Cambridge: Bellknap Press, 1977).

Ford perhaps using this approach more broadly. Recently announced changes at General Motors involving a consolidation of divisions appear to move them closer to the Ford model.

Nissan, one of Japan's Big Two, developed from Yoshisuke Ayukawa's Tobata Cast Iron Company, which bought a manufacturing license of DATSUN (originally spelled DATSON—a "son" of "DAT" model). Ayukawa was much inspired by Henry Ford's mass-production strategies, as manifested in the Model T. He intended to establish a passenger-car-oriented manufacturer and opposed the government's strong push toward government-controlled truck production. His approach was more or less similar to GM in the sense that Ayukawa's Nissan was a management-oriented company with strong financial ties to his Nissan conglomerate. This approach was much influenced by the business successes of the two American subsidiaries, Japan Ford and Japan GM.

On the other hand, Toyota, a newcomer in the late 1930s, was more or less similar to Ford. It had a strong family tie, backed by a couple of good managers from outside. Kiichiro Toyoda, the founder of Toyota Motors, was a graduate of Tokyo Imperial University's mechanical engineering department, like Ayukawa. However, unlike Ayukawa, Toyoda had an engineering rather than management orientation. He himself designed engines. Therefore, though not completely comparable, Toyota at its inception was quite similar to Ford.

Internally, however, Nissan and Toyota have come to have almost similar structures vis-à-vis parts suppliers. The well-known *kanban* or just-in-time system has been adopted by almost all Japanese manufacturers, though its name was conceived first by Toyota. That the Japanese auto manufacturers are primarily assemblers with minimum formal vertical integration is the critical distinguishing organizational characteristic compared to their U.S. counterparts. In the case of General Motors, decentralization is accomplished through their complex divisional structure. In the case of the Japanese auto producers, decentralization is accomplished through their structure of affiliated companies. There tends to be minimum formal reporting to the parent company and minimum staffing on the part of parent companies to supervise these affiliated companies relative to what would be normal reporting and staffing relationships and requirements in the U.S. divisional structure.

Development of Manufacturer-Supplier Relations

The first passenger car to be produced in quantity in the United States was the 1901 Curved Dash Oldsmobile. It was assembled from components provided by suppliers: Leland and Faulconer engines, Dodge transmissions, and Briscoe bodies. Although over the years the U.S. vehicle manufacturers have developed the capacity to produce many of their own components, particularly the engines, transmissions, and bodies, reliance on suppliers is still high and expected to increase as shown in the following preliminary table from the University of Michigan Delphi III study:

Table 1
Projected Future Reliance on Suppliers

| | Percent* of Parts, Components, and Materials Purchased | | |
| | Estimated | Forecast | |
	1983	1987	1992
AMC/Renault	60%	65%	70%
Chrysler	60	65	70
Ford	50	55	60
General Motors	40	50	50
VW of America	55	60	65

*Dollar-volume basis

Obviously, when even General Motors is expected to be purchasing 50 percent of its parts, components, and materials by 1987, and other U.S. vehicle manufacturers (OEMs) a higher proportion, the always important relations between the OEMs and their suppliers are increasing in importance. In the past, OEM-supplier relations in the United States changed frequently and could best be characterized as erratic and inconsistent. The pattern ranged from severe arm's-length bargaining to "cozy arrangements" that did not serve to generate sustained increases in productivity and quality. Now we are witnessing the rapid development of a model that reflects the following characteristics:

1. longer-term relationships
2. fewer suppliers
3. less reluctance to single source
4. more emphasis on stability and reliability of supply
5. more emphasis on quality.

Within this industry-wide pattern, however, there will remain significant differences between companies—in part because of differing management personalities and philosophies, but also because of major structural differences in the manner in which the U.S. companies evolved.

General Motors grew at first through acquisitions, including many that it still holds. Ford expanded internally and at one time seemed destined to produce everything. As shown in table 1, both companies now purchase 40-60 percent of their supplies on a dollar-volume basis, and Ford more than GM. But Ford produces some items it uses—including steel and glass, which GM buys. These and many other differences, such as the major imports from overseas affiliates by AMC and VWofA, affect each company's individual purchasing habits and have a major effect on supplier relations.

In contrast to the American case, one may find the salient difference of Japanese component suppliers in their historical background and relation to OEMs. As widely recognized, Japan's closely coordinated component supply system (often referred to as the *kanban* system in the foreign literature) owes much to the close relationship between an OEM and its special-component suppliers. This structural relation stems from the fact that a majority of the Japanese automotive OEMs developed from nonautomotive manufacturing firms that had a subsidiary structure for component production.

When Japan Ford and Japan GM were established and almost took over the Japanese market for passenger cars in the 1920s, the contribution of Japanese indigenous parts suppliers was almost nil. Later, in recognition of the importance of components of high quality, the government enacted two laws (the Law for Promotion of the Motor Vehicle Manufacturing Industry and the Regulation for Approving Excellent Automotive Parts and Materials) for strengthening the components industry.

In the aftermath of World War II, demand for components from the Occupation Army and the Korean War led to the rejuvenation of the Japanese parts industry and to modest financial success. Later, in conjunction with the Foreign Technical Contract Regulations in 1950, the government promulgated the Temporary Law for Promotion of the Machinery Industry in 1956. Although this law did not specially aim at the automotive-component industry (see chapter 5), it had a strong impact, leading to significant improvement in the quality of auto parts. In the meantime, the Japan Industrial Standard (JIS) was established for motor components under the Industrial Standardization Law of 1951.

It is important to note that independent, qualified parts subcontractors did not exist at the outset of the Japanese automotive industry. Therefore, the OEMs had to take the initiative in developing them within their framework of business relationships. This role contrasts with the way manufacturer-supplier relationships developed in the United States. That is, Japanese manufacturers of motor components have been accustomed to being dependent on the OEMs, financially and technologically. However, important new trends are occurring in this regard. Specifically, there is growing technological and sales independence on the part of component suppliers. This feature will be discussed further in chapter 9.

Technological Change

The history of the automobile industry is one of massive technological change, though the form and character has varied over time. In the early years of the automobile, the architecture (the specific arrangement of design elements) was almost random. Steering, energy sources, and layout varied widely from model to model and year to year. Structural and mechanical principles showed a tremendous range of strategies. This is typical for the early stage of consumer products. A consumer product at a comparable stage of development today would be the personal computer, where a large number of companies compete based on widely varying design strategies.

Over time, as specific products met with success in the mass market, standardization of auto architecture gradually took place. The Model T, introduced in 1905, became the dominant design that was to last for almost twenty years. Incremental change gradually replaced radical technological change.[10] That is to say, change increasingly took place within the confines of the basic configuration of the existing architecture rather than fundamentally altering this architecture. The annual model change symbolized this architectural stabilization. A driver today would have little trouble driving a 1925 Chevrolet.

Abernathy argues that this standardization of product design was particularly pronounced in the United States as a result of the emphasis on meeting consumer preferences in a mass market. He contrasts this with the greater technological diversity among European producers as a result of their more recent history in responding to a high-performance luxury market in segmented national economies.

That post-World War II changes in U.S. product technology have been primarily incremental should not obscure the improvements that have occurred. There has been an interacting set of complex improvements in engine, drivetrains, fuel systems, suspensions, materials, lubricants, etc. These incremental improvements have increased routine maintenance requirements from every 1,000 miles to every 6,000-7,000 miles. Engine life has been increased to 100,000 miles or more before a major overhaul is needed. Safety is another area of major vehicle improvement. The U.S. fatality rate

---

[10]William Abernathy develops this theme in detail in *The Productivity Dilemma* (Baltimore: Johns Hopkins Press, 1978).

per 100 million vehicle miles has dropped from about 12.0 in the pre-World War II period to 3.5 in 1980. Although some of these improvements are due to the improved road system, much is due to the improvement in engines, transmissions, bodies, etc. of U.S. autos and trucks.

Much of Japan's early postwar automotive technology was influenced by the process of technological borrowing. For example, Nissan, Isuzu, and Hino had technology-transfer contracts with British Austin, French Renault, and British Hillman, respectively, in the 1950s. The Japanese licensees absorbed the foreign technology through knockdown assemblies. After the seven-year contractual period, they were in a position to integrate what they had learned with their own development programs and thus introduced their own models. Noteworthy in this regard is that Toyota failed to reach agreement on a technical contract with Ford and went on to develop its own indigenous models. The minicar manufacturers, plus Honda, have consistently shown their unique technological capabilities to produce low-priced but high-performance vehicles.

If we consider the government's past auto policy, we see that the focus was not the automotive OEMs (who were primarily assemblers) *per se*, but the promotion of general technical development of the small and medium machine-tooling companies. In this policy environment, the automobile served as a good policy concept to promote all parts and machine-tooling industries.

In order to give a rough historical sketch of Japan's automotive technology, figure 5 shows that the number of patent applications for transportation machinery (of which the majority is accounted for by the automotive industry) started its upward growth well in advance of other manufacturing industries. This implies that already before 1970, Japan's automotive technology was receiving great stimulation. In fact, the highest increase was attained prior to 1970. The technology for pollution control and fuel efficiency that appeared in the 1970s seemed to have been well-developed by the late 1960s. This also indicates that the Japanese car boom in the world market was partly based on the technological efforts in this earlier period.

One can expect a continuing process of incremental change. There will be a steady refinement of existing design elements. Some major innovations in the future will derive from the marriage of the electronics industry and the auto industry. The continuous variable transmission (CVT) is a specific innovation likely to further advance auto reliability. One of the most difficult technological tasks faced by both U.S. and Japanese designers is to duplicate in today's smaller cars some of the riding comfort and ease of operation inherent in the 300 horsepower and 4,000 pound U.S. cars of the past. The task is difficult because these advantages of larger cars were a function of physics, not design or management strategies.

FIGURE 5

Japanese Patent Applications

(DOMESTIC)

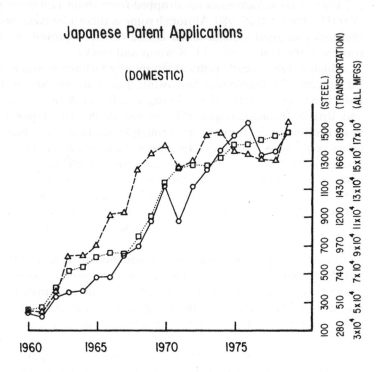

SOURCE: *Tokkyo Koho* [Official gazette]. Tokyo: Patent Office.

o————o **Steel**

△——————△ **Transportation Machinery**

□·····················□ **All Manufacturers**

With the increasing shift to worldwide automotive competition, we can anticipate more diverse sources of product innovations. Yet, through a network of joint ventures and the traditional "public" nature of product innovations, one can also anticipate a continued high diffusion of innovations among the various manufacturers worldwide. Those slower to adopt such innovations will find themselves severely handicapped in the marketplace. It should be noted that, historically, the rate of diffusion of product innovations among producers, such as the automatic transmission and power steering, has been quite rapid.

The process of standardization of design elements was closely related to the evolution of process technology in the industry. Using engine production as an example, Abernathy describes the developments in terms of a shift from ill-structured tasks, highly skilled craftsmen, job-shop work flow, and general-purpose equipment to one of tightly integrated processes utilizing operator skills, "dedicated" equipment, and higher and higher levels of automation. More recent developments involved the increased utilization of automatic transfer machines, which tie operations together without operator intervention, and the development of feedback controls and machine self-correction.

Current trends suggest a modest reversal of these practices with the development of flexible manufacturing systems (FMS). The Japanese auto firms have been particularly notable for the development of production technologies that involve: quick setup of machinery to reduce the costs of short production runs, equipment layout that permits sequential flow of products, multi-skilled operators who can move from one process to another, and an ability to assemble efficiently two or more platforms as well as different models on the same line. These innovations significantly increase the economies associated with small- and medium-lot production. They also increase the flexibility of production facilities in responding to rapidly changing consumer demand, especially in different geographic locations, and make it more economical to respond to small market niches. Just how far these trends can go in reversing past practices remains to be seen, but it seems likely that large production runs will still involve the incremental evolution of the traditional process technologies described above.

Evolution of Government Policy Making and Its Impact on the Auto Industry

Automobiles and government have not had a particularly intimate relationship through most of the postwar period in the United States. However, in comparison with other industrialized nations, it is possible to say that into the 1960s America had an effective, if implicit, "industrial policy" favoring the development of the auto industry. Fuel prices and the various taxes, fees, and credit costs associated with buying and driving a car were kept low. However, vehicle and fuel taxes supported enormous public investment in roads and highways. Tax diversions to competing forms of transport such as intercity and commuter railroads were relatively small. While these "policies" were not the direct result of industry lobbying, automotive interest groups were significant forces in maintaining them.

The net effect was higher demand for automobiles than would be the case if the United States had followed policies common elsewhere, and this demand was predominantly for cars much larger than those produced elsewhere. The American market was thus relatively insulated, though not consciously protected, from the rest of the world. As a consequence, wages and other costs rose and the competitive tone of the industry slackened, but this enormous segment of the American economy prospered and contributed substantially to real growth in productivity and GNP.

Through the 1960s and 1970s conditions in general and government policy in particular became much less favorable to the auto industry. This was a turbulent era of heightened political activism in America, as evidenced by the environmental and consumer movements. The style of politics and many participants changed; a "new breed" of activist congressmen, bureaucrats, and "policy entrepreneurs" gained influence in Washington. Accompanying this trend was a general suspicion of the establishment, including big business. Heightened public consciousness of problems and the new activism that accompanied it led to expanding governmental intervention into many sectors of American society. The auto industry was faced with tough new regulations on auto safety and emissions, along with rules about pollution from their factories, worker health and safety, and so forth that applied to all industry. These came at a difficult time: even prior to the 1973 oil crisis, American consumer preferences were diversifying and competition from imports was rising. The gasoline shortages and price hikes of 1973 (and then of 1979) depressed car sales, increased import competition in small cars, and stimulated more political pressures. For example, despite active industry lobbying and the declared preferences of the political leadership, it proved impossible to move oil prices up to world levels. Cheap gas depressed demand for small cars in the mid-1970s, complicating product decisions by the automakers, and led to a regulatory approach to the "energy crisis" of mandating progressively tougher fuel-economy standards. In the meantime, aggressive enforcement of environmental and safety regulations continued.

The pendulum began to swing back in the late 1970s and accelerated with the election of President Reagan in 1980. Some regulations (i.e., air bags) were relaxed, and the government (prior to President Reagan's election) arranged financial assistance to prevent the bankruptcy of the Chrysler

Corporation. A voluntary quota to provide some protection against imports was negotiated with the Japanese government, and stronger protectionist policies attracted much support in Congress. Tax laws were revised to encourage savings, investment, and industrial research and development. At the same time, the government's tight money policy brought recession and high interest rates, depressing auto sales. Even during the 1983 recovery, persistent high interest rates and the looming budget deficit created concerns about how sustainable the recovery would be. Efforts to roll back anti-pollution regulations were thwarted by the continued strength of environmental concerns among congressmen and the public. While proposals abound for a more positive "industrial policy" to regain international competitiveness in the automotive and other industries, the Reagan administration remains committed to its "small government" position. Whether its proposal for a new Department of International Trade and Industry presages a change in its orientation, or whether the election of a Democratic president in 1984 might bring yet another radical shift in public policy toward industry are major open questions.

Japan's "industrial policy" as applied to automobiles began in 1936, when the Automobile Manufacturing Industry Law, which favored Japanese producers, drove Ford and GM from their dominant positions. In the extraordinary shortages of the early postwar years, however, the auto firms had to struggle for enough government favor to stay in business—public policy was oriented more toward steel, energy, and other "basic" industries. Few at the time forecast the future success of Japanese cars.

It was the Korean War that put the auto industry on its feet: directly through enormous "special procurements" of trucks for the American military; indirectly by stimulating the entire Japanese economy and initiating a long-term trend of expanding demand for automobiles. In the 1950s this demand could not be met by the domestic makers on their own, and the government authorized the assembly of knockdown kits from Europe at the same time that imports of finished vehicles were barred. Foreign exchange controls were used to increase "local content" annually, until by 1960 all production was completely Japanese.

Throughout this period the auto firms benefited from a variety of government supports: long-term loans from public or government-influenced banks, accelerated depreciation and tariff exemptions on needed production machinery, tax breaks for export promotion, and others. In general, these were neither crucially important nor specifically designed for automobiles (although the auto-parts industry was "targetted" for special help from 1956 to 1970). The key government policy was thoroughgoing protection against both imported cars and, particularly, direct foreign investment. Trade barriers had been erected around virtually all Japanese manufacturing for a time, but by the early 1960s MITI officials had become particularly concerned about the vulnerability of the auto industry and preserved its protection until as late as possible.

MITI's main response to the threat of internationalization was an often repeated effort to concentrate the auto industry. In 1955 it had offered a semimonopoly to the first company to develop a cheap "people's car"; in the 1960s it tried to organize the industry into two or three specialized "groups" and allocated special loan funds to encourage mergers. These efforts mostly failed because the auto firms had become powerful enough to resist government advice.

In short, a variety of government programs were helpful to the auto industry during the explosive expansion of domestic sales in the 1960s and exports in the 1970s. However, with the exception of trade protection at the infant and toddler stages, these factors were not as important to the industry's success as the rapid growth of household income in Japan, steadily increasing world trade, skillful and aggressive management, and the "accidents" of the two oil crises in the 1970s.

During this latter period, relations between the government and the auto industry were relatively distant and, in fact, were dominated by environmental and safety controversies quite similar to those in the U.S.: strong social movements pushed the government into passing tough regulations that were strongly opposed by the auto firms. Their resentment at government interference increased when MITI established export restraints in response to pressures from Europe and the United States. Although the Japanese auto industry currently has confidence in its international competitiveness, its growth has been stalled by a sluggish economy at home and strong protectionist trends abroad. If either problem is to be solved, government must play a major role.

Transformation of Labor-Management Relations

The history of labor-management relations in the U.S. automobile industry defies neat encapsulization. Trends in wages and benefits, time lost to strikes and work stoppages, and indices of worker-manager relations (e.g., grievance, absentee, and discipline rates) suggest two things: (1) over time, workers and managers have benefited economically from the market strength of the auto companies, but (2) even in periods of greatest economic prosperity, certain factors have contributed to cleavages between those who owned and managed these enterprises and those who built their products. To summarize the history of labor-management relations as "adversarial" perhaps captures the nature of the tie between employers and workers (i.e., any meeting of buyer and seller in the marketplace presumes an opposition of purposes), but it ignores the broad avenues of agreement that have tied employers and workers to a "community of fate" (e.g., collective-bargaining law and procedures, grievance mechanisms, apprenticeship and training programs, procedures for achieving a mutually acceptable allocation of labor to different jobs and shifts, and, most recently, profit-sharing systems).

Two observations may help to structure our understanding of labor-management relations in the U.S. auto industry. First, the present emphasis on communication and cooperation between unions and management is actually quite consistent with the repertoire of past practices, which produced both consent and conflict in the industry. Alongside the ill will and occasional violence that have accompanied labor-management relations since the 1930s, a framework for cooperation has evolved: the financial well-being of the auto companies has, with the rise of unionism, also provided a foundation for the privileged economic position of auto workers relative to employees in other sectors of the economy. Similarly, the stability of the auto companies as principal suppliers of transportation to an auto-dependent public has resulted in a generally stable demand for labor (though not always in the short term) and, more importantly, a demand for (and rewards to) a stable labor force. In other words, high wages, extensive benefits, stable jobs, a legalistic handling of employment and grievances, and the construction of an elaborate collective-bargaining apparatus were founded upon and strengthened an implicit cooperation between labor and management. This is the case, notwithstanding the political rhetoric engaged in by both parties to the labor market.

Second, the very success of the U.S. industry in creating large enterprises and sustaining their growth has contributed to a form of impersonalization and organizational rigidity on the part of both employers and unions. To the extent that the auto companies historically have attempted to respond to uncertainties in the market by asserting increased control over the processes and personnel involved in production, they have teetered on the brink of overcontrol, which is manifested in allegiance to "established procedures," rigidity in thinking, and a refusal to recognize changing circumstances external to the organization. These practices have stimulated (though not directly determined) a strikingly similar response on the part of unions: unsolved problems of alienation and changing attitudes on the shop floor have tended to draw a response out-of-step with the times—i.e., a belief that greater economic rewards will reelicit the consent of auto workers. And, ironically, the application and elaboration of the traditional instruments of cooperation—higher wages, broader and more complex grievance and seniority procedures—have only acted to further complicate and rigidify the system of labor-management relations.

An outstanding feature of the Japanese automobile manufacturers has been their systematic and intensive effort in developing human resources, particularly those of manual production workers. Orientation, education, job training, and other human resource development strategies for blue-collar workers are essentially comparable to those for white-collar (managerial, clerical, and technical) workers in Japanese auto companies. There are, of course, differences in the intensity and breadth of training, which reflects the differences in their functions. Moreover, in the case of Japanese companies, workshop group activities have played an integral role in fostering skill development in the blue-collar labor force. Participation in small-group problem-solving activities (e.g., Quality Control [QC] circles) has proven an indispensable vehicle in this task. In addition, these activities have been employed to familiarize workers with the organizational environment of the company and people, and even to help assess an individual employee's potential for promotion.

The strong emphasis on human resource development with regard to Japanese production workers reflects the historical conditions of resource endowments and competitive pressure. With a relative shortage of capital and technology and a relatively abundant supply of young and able laborers in the early phases of the industry's development in the postwar period, it was to the advantage of Japanese auto manufacturers to concentrate on the effective utilization of human resources. Investment in human capital was imperative. In retrospect, this appears as one of the most important policy differences between U.S. and Japanese auto companies.

Another important factor that should be borne in mind is the experience of postwar labor reforms in Japan, promoted largely by the Allied Occupation Forces and the vigorous movement of newly organized labor unions. With the advent of the union movement, discriminatory practices that pervaded Japanese companies in the prewar period and that decisively distinguished blue-collar workers from white-collar workers were largely abolished. Among the most notable developments of the enterprise unions containing both white- and blue-collar employees was the unification of pay systems, promotion practices, education and training policies, fringe benefits, in-company social statuses, and other employment conditions. This background is indispensable in understanding the relatively egalitarian treatment of blue-collar workers in the human resource development strategies of the Japanese companies.

After a turbulent decade of labor-management strife following World War II, joint-consultation systems developed as an important vehicle for coordinating management and union activities. By means of extensive and intensive interactions between management and labor side-by-side with collective bargaining, management and unions share common information and jointly seek workable solutions to organizational problems. However, the joint-consultation system is only one element of the complex fabric of information exchange within the corporate organization. Equally important are the formal and informal discussions of managers and workers at the workshop level and the leadership role supplied by first-line supervisors.

The contemporary human resource development strategies of Japanese auto companies are not, however, without problems. In recent years, their weaknesses have grown increasingly visible. This is particularly the case with diminishing prospects for promotion resulting from slower growth in the industry and the aging of the work force. Expectation for promotion has been the single most important source of motivation and commitment for Japanese workers. Recent structural changes have given rise to a new series of challenges to the historically most advantageous aspect of management in Japanese auto companies, as well as for companies in many other Japanese industries.

Conclusion

With this general introduction complete, we are now ready to pursue the specific topics listed in the table of contents. It is clear from our treatment that the historical circumstances and trajectories of the two industries were quite different in a number of critical respects. Yet, both industries have come to assume primary importance to their respective economies. The extent to which they will retain their differences or become more similar is one of the major themes to be pursued in the subsequent chapters.

Chapter 3
Internationalization of the Automobile Industry

Introduction

The world automotive industry is moving through momentous changes, and one of the major changes about which observers speak is the internationalization of the industry. While "everyone knows" what is meant by the internationalization of the auto industry, on closer examination we find that "everyone" seems to know something different. We begin with a "bare bones" definition. Internationalization of the auto industry is the diffusion of production, consumption, and capital investment of finished vehicles and auto components from the established world centers to new centers.

Our research focused, first, on documenting the existence of these processes and the form they have taken and, second, on analyzing their driving forces and consequences—present and future. The broad forces driving internationalization forward over the postwar period derive from a variety of often conflicting factors. Of particular prominence are:

1. progressive liberalization of auto tariffs and other policies inhibiting auto trade by the industrially advanced nations;

2. strengthening of trade barriers in order to further national economic development by many developing countries, as well as selected protectionist actions by industrialized nations to protect balance of payments, industry, markets, and employment;

3. increasing worldwide commonality of consumer desires for similar kinds of passenger cars;

4. growing capital costs associated with designing, producing, and marketing cars and the resultant pressures for cooperation or merger among auto companies;

5. growing competitive pressures that lead auto firms to reduce costs through worldwide sourcing of parts and subassemblies; and

6. efforts by trans-national corporations (TNCs) to establish or secure their positions in world auto markets by locating production facilities closer to growing markets and to take advantage of lower factor costs.

43

We will discuss all these factors in the course of this chapter, but for the moment we focus on the first two factors.

In examining this list, we see first that there are a variety of actors involved: governments, the auto industry, individual manufacturers, consumers, and unions. Even within a particular country, these actors often have conflicting interests and therefore pursue different strategies. We also see that these driving forces represent a variety of crosscurrents. Some currents are consistent with the extension of the liberal trade order (defined as the progressive removal of direct restrictions affecting trade across national boundaries), and others work against it. Moreover, the linkages between the various factors driving internationalization are exceedingly complex. For example, domestic-content legislation, customs, and tax barriers in the developing countries surely run counter to free-trade principles, but, under certain circumstances, they have clearly contributed to an expansion of international trade involving those nations adopting such restrictions. The recent experiences of Spain, Mexico, and Brazil are instructive in this regard. While the barriers to free trade in these countries have contributed to a domestic auto industry and their participation in international trade as exporters, they have done so only in a climate in which other nations had relatively few such barriers. This is a classic case of the "prisoner's dilemma" game in which the cooperating partners (in this case, free-trade countries) provide benefits to the defecting partner (countries erecting trade barriers while exporting). Whether the absence of such restrictions would have contributed still more to increasing international trade is debatable. Nevertheless, these restrictive practices have been compatible, at least in the short run, with an expansion of world trade involving these countries. And expansion of world trade is the presumed core outcome desired by advocates of the liberal trade order. This means that the annual growth of auto trade throughout most of the post-World War II period is not a result only of a strengthening of the liberal trade order.

This seeming anomaly becomes less confusing if we disaggregate aspects of international trade over the last two decades. Based on IMF and GATT data, Bo Ekman,[11] Senior Vice-President of Volvo Corporation, estimates the "free" portion of total international trade at 25 percent in 1981, compared to some 66 percent in early 1960. The remainder in 1981 was accounted for by barter (25 percent), quotas and orderly marketing agreements (25 percent), and internal trans-national corporate trade across national boundaries (25 percent). These estimates find support in research carried out independently by the staff of the *Economist*. One can debate what constitutes the "free"

---

[11]Bo Ekman, "To Get Sweden Going: Thoughts on Industrial Policy and Corporate Strategy" (Volvo Corporation Internal Document, 1982).

portion of trade, but clearly there has been a gradual movement away from free trade, especially accelerated by the political restrictions imposed on trade in the early 1980s.

It is apparent that the phrase "liberal trade order" has always masked a reality of much greater complexity. The original GATT regulations negotiated in 1947 represented a delicate balance between the needs of national governments to carry out their domestic obligations and the desire to remove the worst obstacles obstructing international trade. The current upsurge of protectionist sentiment must be viewed against its already high and well-developed level. From the point of view of many developing nations, the liberal trade order best served the interests of the industrially advanced nations, and, therefore, the initiation of protectionist policies by developing countries were simply efforts to provide momentum for their own national growth. We see, therefore, that not only does the definition of internationalization vary according to one's focus, but the very benefits to be accrued from internationalization and the liberal trade order are evaluated quite differently by different parties at different economic stages.

It follows from this discussion that the definition of internationalization and its perceived benefits often varies according to corporate and national interests and may change over time. Our task is to treat these varied aspects of internationalization primarily as context for the specific analysis of the relationship of the U.S.-Japan auto industry. As the two largest auto-producing and auto-consuming nations, it is only natural that discussions of internationalization should focus on these two key actors.

The U.S. and Japan Perspectives

With these considerations in mind, we turn first to the American case. The once dominant and largely captive domestic large-car American market has dramatically shrunk in the face of demand for small, fuel-efficient cars. Notwithstanding the uncertainty of energy prospects, no serious observer anticipates a return to dominance of the old market for full-size cars (see chapter 6).

The significance of the demise of the large-car market in the United States is twofold. On the one hand, the convergence of consumer desires for similar kinds of automobiles in the U.S. and non-U.S. markets provides an enormous market for imports of components and finished vehicles into the United States. On the other hand, this convergence makes possible for American manufacturers, on an unprecedented scale, the integration and coordination of a strategy for worldwide product development (engineering and design), production, and marketing. For a variety of competitive and historical reasons to be discussed below, internationalization does not seem to provide significant new opportunities for U.S. manufacturers to export components and finished vehicles directly from the United States, either to Japan or to other foreign markets. Given the relative importance of the United States—as both a producer and consumer of finished vehicles—the implications of these opportunities and constraints for world trade are of major importance. These are the specific forces driving internationalization of the auto industry from the U.S. perspective.

From the Japanese perspective, there have developed unprecedented opportunities for the export of components and finished vehicles throughout the world. However, the importation of finished vehicles and auto components into Japan is miniscule.

Because of these discrepancies in trading patterns between Japan and other nations, there has arisen enormous tension between Japan and its trading partners. This, in turn, has led to a variety of restrictive actions against the import of Japanese components and finished vehicles. Japanese finished passenger-car imports were restricted in 1982 in markets that account for about 77 percent of the non-Communist world demand outside Japan (up from 20 percent in 1980). Moreover, this does not include those developing countries with tariff barriers and domestic-content requirements not specific to given nations. This worldwide resistance to Japanese imports threatens the heretofore successful Japanese strategy of exporting finished vehicles from their homeland base. In response, Japanese manufacturers are reevaluating their strategy.

What are the options open to Japanese producers as they seek to restore a growth trajectory for their industry given weak worldwide growth in auto demand, especially in the industrialized nations, and given the constraints arising from protectionist sentiments? The following are the emergent strategies:

1. Expand market share in the nonindustrialized nations experiencing more rapid growth (often at the expense of Western producers).

2. Locate production facilities abroad, preferably knockdown assembly operations that preserve high value-added operations in Japan, but be prepared to build higher local content into those operations that are in highly strategic markets should political conditions require it.

3. Enter into cooperative ventures with Western producers, including the establishment of local production, in order to secure access to their domestic markets. Enter into cooperative ventures in Third World nations both to gain access to important potential markets as well as to engage in "production sharing" in a way that Japan gains the benefits accruing from abundant labor at lower wage levels.

4. Increase their role as sources for parts and subassemblies used in the products of Western auto producers.

5. Move to "upscale products" in those markets where quantitative limits have been established on their imports.

6. Use "third" nation production as "staging areas" to gain access to desirable markets, such as represented by the Ford-Tōyō Kōgyō agreement on a cooperative effort in Mexico that will include exports to the U.S. and Canadian market and the use of Spain as well as other individual countries by Japanese and U.S. producers to gain access to EC markets.

Few of these strategies are mutually exclusive, and individual Japanese producers are pursuing them with varying emphases. Nissan Motor Company, for example, which has been quite active in forming joint ventures throughout the world, envisions by the end of the decade selling one-third of its production domestically and one-third through exports and assembling one-third abroad. What the precise mix will be between these various strategies will vary by opportunity and individual corporate situations and initiatives. Although it remains to be seen how adequate these strategies will be to achieving the aims of Japanese producers, they are moving more rapidly in these directions than is commonly recognized, and these activities represent different facets of the internationalization process.

Disaggregation and Diffusion of Auto Production, Assembly, and Capital

Worldwide automotive production and consumption, viewed from the historical perspective, have grown significantly over the past thirty years. The growth has been accompanied by a diversification of sites of production and centers of automotive investment, purchase, and use. Car production (relying wholly or mainly on domestically produced parts and components) took place in twenty-four countries in 1981. Some 5.0 percent of total auto production took place in developing countries; this was an increase from 2.0 percent in 1960. Mexico, Brazil, and Argentina alone accounted (in rather equal shares) for some 91 percent of the developing-nation total.

When we turn to examine the total CKD estimated assembly (relying wholly or primarily on imported parts and components) of 2.6 million vehicles in 1980, the industrially advanced nations accounted for 55.3 percent. The developing nations accounted for the remaining 44.7 percent, with some thirty countries having motor-vehicle assembly operations. The share of assembly operations by developing nations is estimated to have risen from 20 percent in 1967 to its current level of 44.7 percent. The leading CKD assemblers in 1980 were: Nigeria, South Africa, Venezuela, and Iran. In summary, the greatest diffusion in national sites has been in CKD assembly operations, with significant, though less rapid, diffusion occurring in the complete production of automobiles. It is in the latter category that one would expect to find "new contenders."

Finally, the least diffusion has been in national sources of capital. This is not surprising since much of the diffusion of production and assembly sites has been accomplished by the TNCs. Moreover, with the trend toward consolidation and rationalization of existing producers, the prospects for diffusion of national sources of equity seem difficult at best. What shows up are, on the one hand, rapid increases in manufacturing exports from given nations and, on the other hand, the process of reshaping and relocating the chain of production facilities by TNCs. The TNCs have accepted some degree of local equity ownership in marketing networks and assembly operations.

Two further observations should be made. First, international trade in finished vehicles and, recently, components has played an important role in this diffusion process. The North American share of the world market has declined as first European and then Japanese shares grew rapidly. Similarly, the share of production in non-Communist nations other than the United States, the EC, and Japan grew to 17.4 percent by 1980. Just as it was difficult to predict Japan's rise to international prominence in the industry just twenty years ago, it remains to be seen what new entries will alter the current calculus.

Corporate and National Strategies

A wide variety of strategies has been used by automakers in organizing to serve these global markets. These range from a policy of producing in one's home market vehicles suitable for use in that home market—and, with some adaptation, for export to large parts of the world—to one of directly investing in production facilities worldwide, with vehicles being designed with common and specialized needs in mind. Japanese producers, for the most part, have opted for production in their home market and export abroad, while U.S. manufacturers have stressed direct investment in production facilities worldwide. It is important to keep in mind, however, that there are significant intranational variations in corporate strategy. Moreover, as we have seen, there are alternatives within this range, such as the use of third nations as "staging areas" for imports into other countries. American producers have often justified their strategy of direct investment by stressing the importance of "providing jobs and paying local taxes in the markets they serve." Japanese producers similarly have stressed how their exports of low-cost, high-quality products best enables them to contribute to consumers in the various world markets they serve. Here, we note only that this difference in strategy might be accounted for by the strength of the dollar as a key currency and Japanese government control of foreign investment throughout most of the postwar period.

International Trade in Finished Vehicles

We turn now to the extent and form of international trade in motor vehicles, looking initially at direct exports. Figure 6 summarizes the growth in exports for selected countries since 1960. Japan, of course, shows the most dramatic growth in exports, surpassing the French and Germans in the mid-1970s. German automotive exports have been relatively stable since the early 1970s, while French automotive exports have shown significant growth, moving from some 1.0 million units in 1968 to 1.9 million units in 1982. For the United States, once we net out exports to Canada, we see a relatively inconsequential level of exports.

If we compare aggregate Western European trading patterns with Japan and the United States, it is clear that over the last decade Japan has moved dramatically upwards to share with the Western European nations the common feature of having an export ratio of roughly 50 percent. The U.S. performance is striking, with only a 10 percent export ratio. Furthermore, once we exclude exports to Canada, it drops to only 2 percent. Yet, there are also some striking differences between Japan and Western European trading patterns. Whereas France and Germany, for example, both export some 50 percent of their production, they also import some 30 percent of their registered cars. The ratio of import to domestic sales is even higher for Italy and the United Kingdom (45 and 60 percent, respectively). This outcome reflects the substantial trade within the Common Market. Yet a parallel situation does not exist in the case of Japan, where imported vehicles have never exceeded more than 2 percent of annual registrations in the postwar period.

Several further observations are in order regarding the growth of trade over the past decade. First, the volume of exports grew at a compound annual rate of 5.3 percent. World trade in motor vehicles has been a significant and growing factor for many years. Second, it is Japan that has most significantly increased its share of total world exports over the last decade. Third, West Germany has experienced the sharpest decline, with the United States experiencing a modest increase from 4.4 percent in 1970 to 5.3 percent in 1980. By 1981, as noted in chapter 2, Japan accounted for 31.9 percent of world trade in motor vehicles, compared to the 7.2 percent figure for the United States.[12]

---

[12]U.S. export totals include shipments from the United States to Canada.

FIGURE 6

Motor-Vehicle Exports: Selected Countries

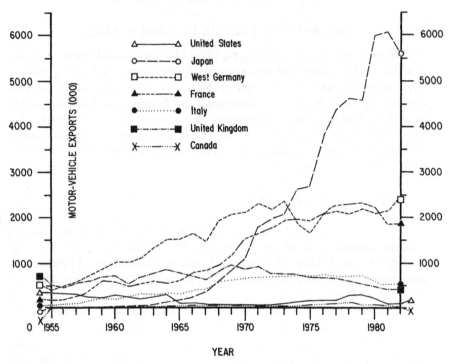

SOURCE: *World Motor Vehicle Data*, 1976-1983 Annual Issues, ISSN 0085-8307. U.S.- Canada trade is not included in their respective statistics.

Table 2 provides a snapshot of world trade in *passenger cars* for 1981. Excluding intra-European and U.S.-Canadian trade, the Japanese accounted for 74 percent of world trade in automobiles by the end of 1981, with North America accounting for only 2.1 percent. It may be seen that 52 percent of Japan's exports are to North America. In recent years, most of its other export markets, such as Southeast Asia and Europe, have been growing at a faster rate than the North American market. By contrast, we see in table 2 that one-quarter of the small total of North American autos go to Latin America. Further examination of table 2 shows that Europe was barely a net exporter of passenger cars in 1981, culminating a long period of decline as its exports failed to grow commensurate with the growth in imports. Only Japan of all the world areas enjoys a healthy surplus of passenger-car exports over imports.

Thus far, our discussion of exports has been geared to a comparison of direct exports, and it is clear that the United States has, by and large, not participated in the growth of direct export trade in autos. Nonetheless, as discussed in chapter 2, this presents only a partial picture, since an alternative is to develop subsidiaries in foreign nations as a means of participating in the growth of their markets. General Motors and Ford have established subsidiaries in a number of countries to insure participation in their domestic markets. Notable, of course, is the relative lack of participation of the Japanese in this strategy. Ninety-four percent of Toyota's production, for example, takes place within the confines of the Japanese mainland.

Obviously, the strategic choice of whether one engages in direct export from a home base or one uses subsidiaries for export has major economic consequences. Most apparent are the varying impacts on employment levels and balance of payments between nations. In summary, although the principal U.S. firms have not participated in the direct export strategy to any appreciable extent, they have benefited considerably from the growth in world auto markets through the establishment of large numbers of productive bases in other nations. This allowed them both to participate in the growth of these domestic markets and to use several of these nations as staging locations for exports of finished vehicles and parts.

New Entries

Barriers to entry in the auto industry—deriving from high economies of scale (with associated high capital costs) and currently high rates of product and process innovation requiring strong technological expertise—make it unlikely that there will be a "new Japan" in the next decade. The recent apparent decision by the Koreans and the Taiwanese, who had modeled their strategy heavily on the Japanese, to ally themselves more closely with the

Table 2
1981 World Trade in Passenger Cars
(in thousands of units)

		Exports By:						
		North America	West Europe	Japan	Latin America	COMECON	Rest of World	Import Total
Exports To:	North America	X	553	2116	0	16	0	2686*
	West Europe[1]	14	X	943	42	137	2	1137*
	Japan	4	27	X	0	0	0	32*
	Latin America	29	131	204	X	1	0	366*
	COMECON	0	15	7	0	X	0	22
	Rest of World	71	417	831	0	0	X	1320*
	Export Total	119*	1144*	4100*	42	154	3*	5562*

*Totals fail to sum due to rounding error (+/- 1000).

[1]For Latin America, COMECON, and Rest of World, 1980 data for the United Kingdom is substituted since a Customs and Excise strike during 1981 prevented collection of 1981 data. For North America/Japan to/from the United Kingdom, 1981 data was obtained from North American and Japanese sources.

NOTE: Wherever the data permitted, knockdown assemblies were excluded but station wagons were included.

Sources: Motor Vehicle Manufacturers Association of the United States, *World Motor Vehicle Data*, 1982 edition, *Motor Vehicle Facts & Figures '82*, *U.S. Imports of Motor Vehicles and Motor Vehicle Parts*, and *U.S. Motor Vehicles and Motor Vehicle Parts Exports*. See also *The Motor Industry of Great Britain, 1981*, *Statistics Canada, 1982 World Automotive Market*, and *Motor Vehicle Statistics of Japan, 1982*.

TNCs suggests the difficulty of going it alone under the new conditions of the industry. The planned state economies of Eastern Europe and the Soviet Union and China represent still a different potential. The centrally planned economies of the Soviet Union and Eastern Europe accounted for 8 percent of total automobile production in 1981. Exports by Eastern bloc countries to Western Europe increased significantly in the past decade, but political considerations and low quality levels in their consumer-goods industries seem likely to set limits to their growth in the future. Rather than a "new Japan," it seems more likely that the trans-national production and trade of auto parts rather than finished vehicles will provide specific niches for a number of the newly emerging nations. This "disaggregation" of auto production and trade from finished vehicles to auto parts and technology may further diffuse production sites, satisfy specific concerns over national balance of payments, and provide a multi-layered stability to trading relationships.

Yet, it still cannot be ruled out that specific countries that build up their expertise, economies of scale, and infrastructure within the framework of the TNCs might not try to strike out on their own in the future. Strong nationalistic sentiments push many in this direction. One official of the U.S. auto industry estimated that developing auto producers would need to depend on technology from the TNCs until they could produce between 250,000-300,000 cars a year. With a well-developed component infrastructure, the threshhold could be lower. Such an outcome, should it occur, does not appear likely to have a direct impact on the markets of the industrialized nations within the next five years. However, the Koreans and Taiwanese, in particular, may well take a significant market share in the developing nations thereafter. In collaboration with selected TNCs, assuming political conditions permit, they could also take a market share in the Western industrial nations. Early 1984 announcements suggest that rapid changes may be forthcoming. These announcements include Hyundai's (Korea's largest automaker) plans to begin exporting cars to Canada and GM's tentative agreement with Daewoo Corporation (in which it owns 50 percent interest) to jointly produce a small car in Korea, with up to half of the vehicles available for export to other countries, including the U.S.

Future Worldwide Auto Demand and Capacity

As will be discussed in greater detail in chapter 6, it is generally agreed that the overall growth in consumption of motor vehicles will be lower in the foreseeable future than in the recent past. An annual, compound growth rate worldwide of about 2-4 percent provides a generally accepted basis for discussion. Certain markets have been expected to grow faster than this average rate (e.g., Latin America, Eastern Europe, and Southeast Asia). The rate of per capita income growth and government policies will determine the rate of growth in auto demand. Generally, these are areas where per capita income is just getting to the point where personal motorized transportation is a realistic objective for significant population groups. However, political risks are often high in such countries, and their ability to maintain momentum in economic growth rates is problematic; the recent experiences of Poland, Mexico, and Argentina are painfully instructive in this regard.

As will be discussed in the subsequent chapter, there is and will be an overabundance of auto-building capacity throughout the world if current estimates of demand growth rates and production plans are correct. Pressures to have the Japanese build production facilities in the worldwide markets they serve, as well as the ongoing efforts of Western manufacturers to increase production in selected Third World nations, may lead to further increases in capacity. Moreover, the acceleration of technological innovation leads to further replacement of manufacturing equipment and obsolete factories. These developments indicate a scrapping of considerable production capacity, especially in those facilities producing vehicles and parts not meeting shifting consumer preferences. This, in turn, results in a significant relocation of the labor force, thereby exacerbating political tensions. If strictly economic factors were all that needed to be considered, the Japanese, with their low-cost production bases, would emerge with the most modern facilities and greatest capacity utilization relative to other producing nations. But this assumes that the political tensions can be managed, something that can hardly be taken as given.

Effects of Overcapacity

One of the effects of this world overcapacity, the rapidly increasing development costs for new products, and the uncertainty of floating currency rates has been and will continue to be an acceleration of business risks. A major ongoing response to this situation is the continuing rationalization of and cooperative ventures between auto manufacturers worldwide. In the case of Europe, strong pressures in this direction derive from its historically more fragmented industry than either the United States or Japan. Although cooperative activities in Europe have sometimes taken the form of acquiring equity in other auto companies, the most common form has been to collaborate on product-development efforts and the joint production of components. In the case of the major U.S. manufacturers, the form of cooperation has generally involved taking small equity positions in Japanese firms to produce cars in Japan that will be marketed in the United States and to buy subassemblies from Japan.

Many of the global firms that heretofore avoided cooperative ventures in major markets are now seriously considering them. The fact that GM and Toyota have reached an agreement suggests the strength of the driving forces. In many cases, these cooperative agreements forge strong mutual relationships in the production of common components while allowing vigorous competition in the sale of finished vehicles. Rather than an acceleration of mergers among auto companies as some have predicted, we anticipate a variety of innovative cooperative relationships at the design, production, and marketing phases.

Another potential effect of this short- and long-term overcapacity in the industry is localized or even global trade conflict in automobiles and their components. Everyone appreciates the risks of such moves, including the likelihood, because of the automotive sector's size and strategic importance to national economies, of this degenerating into a broad-based economic war. Nevertheless, economic and political pressures throughout the world have become so strong in this regard that the worst may still happen.

Worldwide Coordination and Integration

The overall trends described in the introduction have made the products demanded in the three primary auto markets of today (North America, Western Europe, and Japan) far more similar than they ever have been in the past. This, in turn, provides a set of opportunities and threats for both producers of autos and component and materials suppliers. In response to this situation, U.S. and some Western European auto manufacturers and suppliers are developing a global strategy involving the worldwide integration and coordination of product development, production, and marketing. For GM, these developments were symbolized in the "world-car" concept. For the European producers, in particular, this has meant forging cooperative relationships with other auto-manufacturing firms. These efforts represent an attempt on the part of all concerned to share the enormous costs of product development. They are also intended to increase the rate of return of the costs and capital investments necessitated by market-driven product changes by spreading them over as large a production volume (cumulatively) as possible with near optimum costs.

Implementation of these concepts varies by nation and, indeed, by manufacturer depending on their market position. It is now widely recognized that a variety of national differences prevent the design, engineering, production, and marketing of a world car. But as an ideal type that drives forward the implementation of guiding principles, the world-car concept provides us with important information about the direction of developments. In particular, it points us toward a focus on international assembly. This, in turn, focuses producers on the expansion of common functional parts and the worldwide sourcing of components. Underlying these features is a marketing strategy that involves the development of designs that are generally acceptable, at least in terms of major subassemblies, throughout the world so that the manufacturers can benefit from longer product life cycles. This marketing strategy applies just as well to the Japanese, though for them the path for achieving it, up to now, has lain in developing cars that are sourced and assembled in Japan but marketed throughout the world by virtue of their universal appeal.

The strategy of engineering and production relating to the world car is seen in GM's strategy for the J car. The objective was to design, engineer, and manufacture the "same" vehicle with global sourcing. This included engines from Brazil to the United States and Europe, and from Australia to Europe and South Africa; transaxles from Japan to Europe, South Africa, and Australia; and five-speed manual transmissions and diesel engines from Japan to the United States. At the same time, the companies' manufacturing units in each respective country would produce most of the specialized parts needed for their own models. It is only certain key components that are produced in specialized locations and at an optimum volume to achieve economies of scale. The engineering objective is not commonality of parts in every vehicle but *compatibility*. With compatibility, the vehicles are designed to accept different sets of components. Thus, most J car models can accommodate either a West German steering column or one from the United States.

The big savings to be achieved from developing cars on a worldwide scale and sourcing them internationally are derived from the economies of scale achieved in major components. Bernard Vernier-Pelliez, the former chairman of Renault, captured this well in his statement made in 1979: "There are no world cars, only world components to be made on a vast scale."

This objective must be tempered with the recognition of the business risk inherent in multi-national strategies of this sort. Sudden changes in consumer taste and/or government policies combined with rapid fluctuations in currency value raise the stakes enormously for those who engage in these strategies. These growing risks, especially pronounced in the rapidly growing "fringe markets," further fuel the various forms of cooperation and merger among the auto manufacturers.

Although these remarks apply in varying degrees to Japanese manufacturers as well, it must be kept in mind that their direct export strategy to date assumes primarily home-based production and export. This means that integration and coordination represent a very different set of problems. It is not surprising, therefore, that the Japanese manufacturers have been reluctant to move offshore. Yet, increasingly they are moving to do just that.

Worldwide Sourcing

An important element of the global strategy of American and some European firms is worldwide sourcing of purchased components and subsystems. Combined with the economic incentives that derive particularly from lower costs (labor and materials) in developing countries and the attractiveness of their growth markets, there are significant political incentives deriving from domestic-content regulations, high tariffs, prohibitions on the import of finished vehicles, and government subsidies.

At another level, integrated car design and global sourcing allow firms greater flexibility in allocating component manufacture among various plants. Particularly in the area of design, these developments are facilitated by dramatic improvements in telecommunications technology. These developments make possible centralized information banks and design centers that draw information from throughout the world. They also will make it possible for an engineer in Tokyo and an engineer in Detroit to simultaneously examine the same drawing and discuss it while sitting in their respective offices. These breakthroughs in telecommunications are also compatible with a variety of business strategies. They can facilitate the extension and strengthening on an international basis of the classic domestic organizational paradigm developed by Alfred Sloan, Jr., in the 1920s. That is to say, these developments allow centralized policy-making groups to more easily co-ordinate worldwide corporate policy while decentralized national operations serve to carry this policy out. Alternatively, in principle, the new technologies can serve just as well to make more effective the direct export strategy. The important point to stress is that these outcomes will be determined not by the technology itself but by the uses of these technologies by specific companies in the context of business strategies. These business strategies will be devised under uncertain and rapidly changing political and market environments.

The Japanese have, by and large, eschewed worldwide sourcing, and the recent modest increases are a response to outside pressures rather than a desired outcome on the part of the manufacturers. These policies stem from the economies the Japanese manufacturers have been able to achieve with their highly integrated domestic supplier system. As noted in chapter 2, some version of the just-in-time system is practiced by all Japanese auto manufacturers. The "flip side" of this strategy once again is the political reaction engendered by huge surpluses in Japan's auto trade vis-à-vis its trading partners and the concentrated effects on auto employment. To put the matter differently, the business risks associated with the American strategy of worldwide design and production must be weighed against the political risks associated with the Japanese approach. In all likelihood, the lines between these alternative strategies will blur as U.S. manufacturers push forward flexible manufacturing systems and the Japanese respond to the political dangers with more worldwide production.

Notwithstanding, we can anticipate only modest annual increases in Japan's offshore sourcing from the existing low base. With regard to the Americans and Europeans, a strong movement is already under way to increase worldwide sourcing, and it can be expected to continue as firms seek to capture lower labor and material costs. Yet, a variety of economic, political, and technical constraints limit the rate and extent of growth in worldwide sourcing, notwithstanding the current fanfare. There are significant logistical risks associated with long geographical distances, low labor quality, and unstable labor relations in many of the developing nations. Sourcing from abroad is inconsistent in many respects with the worldwide movement toward adopting the Japanese style just-in-time delivery system with all the associated economies involved in reducing inventories. Furthermore, with the shift to more capital-intensive operations, the incentive to locate facilities in developing countries because of their low wages diminishes. This varies, of course, by product. Those products that are highly labor intensive, use indigenously produced materials, and consume relatively little energy will continue to be attractive possibilities for sourcing abroad. Naturally, freight costs are a consideration; however, with the exception of certain items of low value per cubic meter, freight cost is rarely a major determinant. A more important variable, as mentioned above, is the potential inflexibility created by the increased "float"; this inflexibility is greater with some products than others.

In addition, the recent rapid changes in product and process technology in the auto industry mean that there needs to be closer cooperation between R&D and engineering staffs, on the one hand, and production operations, on the other. Notwithstanding improved international communications for the foreseeable future, this is still easier to achieve by concentrating such facilities in the industrialized nations.

The political instability of many developing nations further limits a rapid shift abroad. The business risks are great in rapidly shifting situations. Sudden shifts in trade policy can destroy the best-laid plans. As part of its antiinflationary program, for example, the Mexican government imposed quotas that drastically reduced 1982 auto production below the 1981 totals. In addition, an import ban was placed on many optional-equipment components in 1982, and the portion of total Mexican content by value was increased to 60 percent for cars and as high as 80 percent for trucks. During the summer of 1982, chaos reigned in the Mexican economy as the peso collapsed under the weight of escalating foreign debt and banks were nationalized. Accelerating inflation and the new two-tier exchange system further increased uncertainty. Under such rapidly changing and unpredictable conditions, talking in the abstract about worldwide optimum-size component plants reaping huge economies of scale has a fairyland quality.

In summary, although there are significant incentives for U.S., European, and Japanese manufacturers to shift more to offshore production, there are powerful constraints as well. These constraints are likely to temper a drastic movement abroad, particularly to developing countries, which will limit the internationalization of the auto industry as we have defined the term. Yet, the long-term trend seems likely to continue to move in the direction of worldwide sourcing, even if not at the rate that some of the industry leaders predict (or threaten).

Conclusion

With these and other developments we have described, it is clear that a continuing transformation of the worldwide auto industry is in the making. Worldwide coordination and integration both within firms and among firms is the emergent pattern. Yet, at the same time, at least for the next decade, this would appear to be associated with intense competition among the various and shifting coalitions of auto producers as they seek to establish their hegemony in the various worldwide markets.

Chapter 4
Macroeconomic Issues and
Bilateral Trade Relations

Introduction

While many problems of the auto industry emanate from characteristics internal to individual firms in the industry, it is also quite clear that macroeconomic factors play a major role in determining the competitive strength of individual firms and industries. In this chapter, we will describe first the nature of some of the key macroeconomic factors and discuss the options open to governments as they seek to grapple with auto-trade issues. Next, we will turn to short-term trade politics and see how they have operated to produce specific outcomes. Finally, we will treat the emotional issue of the access of foreign cars to the Japanese market. It has been an issue that has been used to justify a number of policy actions.

There is a strong tendency for public and industry policy makers in dealing with trade problems to wish that necessary adjustments be undertaken entirely by other parties. In terms of the diagram below, the parties in each corner want the parties in the other corners (especially the most remote one) to make the accommodations that will ameliorate the situation.

| U.S. Government | Japanese Government |
| U.S. Auto Industry | Japanese Auto Industry |

This preference leads to acrimonious negotiations, pressures, and irritations in trade relations. Productive solutions are more likely to be arrived at with Japanese industry and government policy makers concentrating on developing solutions through Japanese actions, where they have more control over outcomes. Similarly, American policy makers will be most productive in developing solutions through U.S. actions. To be sure, some problems require joint action, and successful outcomes require the good-faith efforts of all parties.

The essence of the problem and the source of the international frictions is to be found in the excess of worldwide manufacturing capacity. This excess capacity derives from the maturing of demand in the major producing countries, the arrival of new capacity in the newly industrializing countries, and, perhaps most importantly, the world economic stagflation and, more recently, recession.

The U.S. Department of Transportation[13] estimates that worldwide excess auto capacity was on the order of 12 million units in 1981, or 24 percent of worldwide capacity. Even with economic recovery, they estimate that the potential worldwide excess capacity in 1985 will be 9 million units. This suggests a capacity-utilization rate of 83 percent. More importantly, the world's capacity-utilization rate is quite unbalanced, with most of the excess capacity in Europe and, especially, the United States. Between 1978 and 1982 there has already been a 16 percent drop in the final-assembly capacity of the U.S. passenger-car industry (see table 15 in chapter 8 for data).

In the economist's simplest models—with all firms having identical costs and easy entry or exit from the industry—a simple change in demand would be shared prorata among all firms or would lead to the creation or extinction of firms. The problem is that, in reality, auto firms have different costs, as do national auto industries. As a result, cutbacks in production will not be shared prorata among national industries, and within these, not between firms. In the absence of government intervention, then, the incidence of capacity reduction may be quite concentrated. Furthermore, adjustment is not costless, i.e., there may well be substantial losses to workers, managers, and stockholders of the affected firms. Finally, it is not even clear, given the possibility of subsidies (whether aimed at shifting adjustment costs on others or not), that it will be the least efficient firms and national industries that will be forced to reduce their output or even to disappear. In any case, overcapacity can be expected to lead to shifts in market shares between national industries.

The employment effects of the weakened U.S. competitive position in the world automotive industry over the last several years have generally and for obvious reasons drawn the greatest attention. There are, however, no systematic national-level statistics that enable one to clearly allocate the recent auto unemployment to such factors as the inappropriate product mix of the U.S. auto producers, the effects of the recession, or the surge of price- and quality-competitive foreign imports. Notwithstanding, it is clear that the resultant unemployment problems have been severe.

A recent University of Michigan survey of displaced workers describes a key part of the problem.[14] Based on a large representative sample from a population of 5 percent of the total number of laid-off auto workers still

---

[13]United States Department of Transportation, *The U.S. Automobile Industry, 1982* (Washington, DC: Transportation Systems Center, 1983).

[14]Jeanne Gordus et al., *Labor Force Status, Program Participation, and Economic Adjustment of Displaced Auto Workers: An Interim Report* (Ann Arbor: University of Michigan, 1983).

eligible for recall as of April 1983, the survey found that only 33.5 percent had been recalled to work at their previous auto-manufacturing employer by the fall of 1983. Twenty-five percent had taken employment in non-auto-related firms, often at reduced wage levels, and 31.5 percent continued to look for employment. Some 10 percent were neither working nor looking for work. Similar research in previous downturns typically reports 70 percent or more of laid-off workers returning to their previous employer in the recovery period.[15] Not only is the percentage markedly lower in this current upturn, but the true return rate is even lower since the data exclude those who have lost their recall rights (commonly, those whose length of unemployment exceeds their seniority). The new pattern has obvious implications for the welfare of auto workers, their communities, and their union.

The fundamental justification for free trade is that the resulting gains in efficiency enhance consumer welfare. While obstacles to free trade are generally undesirable, such restraints have nevertheless been practiced by many countries for many years. Governments can and will resort to protection to overcome problems threatening their national interest or that of powerful groups in their countries. The emphasis, then, should be on guiding any resort to protection and taking such action as will minimize the problems and their costs to the world system, consumers, and producers.

In the next section, we will discuss the sources of the trade problems, distinguishing between both the macro- and microeconomic factors and the supply and demand factors. The succeeding sections, in turn, discuss the political economy of free trade, including the aims of trade policy and the instruments that are commonly used.

Sources of Trade Problems

There is no one cause of the current trade problems. Many factors have contributed, and the relative importance of each is uncertain. Furthermore, macroeconomic problems and the inherent problems of the auto industry itself must be clearly distinguished. The problem of the U.S. auto industry is that so many short- and long-run factors have all come together at once, resulting in an overload.

One must note that solutions to the macroeconomic problems will not in themselves solve the auto industry's problems, nor will solutions to the problems intrinsic to the industry alone solve the problems. It is vital to remember that both industry-specific and macroeconomic problems must be dealt with at the same time.

---

[15]T. A. Corson et al., *Survey of Trade Adjustment Assistance Recipients,* Final Report, Mathematica Policy Research (Princeton, NJ: Mathematica Policy Research, 1979).

In the remainder of this section, we will discuss the problems under the rubrics of macro- and microeconomic and supply and demand issues. By supply issues, we mean those that affect costs, and by demand, those that affect the latter's extent and composition.

Macroeconomic Issues: Supply

Exchange Rates

Most serious observers believe that the dollar is overvalued vis-à-vis the yen (and other currencies), though there is no evidence to support the view that the exchange rate is being manipulated. While a persisting maladjustment in the level of rates is the most fundamental trade problem, it is the *variability* of exchange rates that makes planning difficult. While purchasing power parity may hold in the long run (say twenty years), in the short run deviations may be substantial and persist for long enough to cause serious economic harm. In every year since 1976, the yen-dollar exchange rate has oscillated in a 25 percent range. The swings impose adjustment costs that are, at best, only partially offsetting.

Even if a solution to the variability of the exchange rates is found, this does not mean that the resultant level will be satisfactory. It is possible that because of the nature of the economies involved, the exchange rate that balances the total supply and demand for the relevant monies, including financial flows, may nevertheless pose problems for trade, or free trade, in particular sectors. In addition, the exchange-rate issue is not solely a bilateral one. Final products incorporate raw materials, components, and other inputs from a wide variety of countries, meaning that a number of exchange rates play some role in determining final prices.

In the past eight or nine years, official agencies have intervened extensively in the foreign exchange markets. One can make a good case for intervention in the event of economic or political crises in which the credibility of the authorities themselves has come into serious question. However, the evidence on the utility of the more normal "leaning against the wind" type of intervention is ambiguous. The governments of the major developed economies should be encouraged to coordinate their monetary policies to avoid the overshooting phenomenon of changes in the exchange rate. Because long-run exchange-rate levels are the consequences of so many aspects of the relevant national economies, it is difficult for governments to deal effectively with the problem. National governments have been unwilling to sacrifice domestic policy goals to achieve improvements in those levels.

Further liberalization of Japanese financial markets and of capital flows would help in the long run by creating a demand for the yen as an international currency and by reducing international disparities in real interest rates, which express themselves in trade flows. This policy should become more palatable over time as the U.S. economy becomes a smaller proportion of the total world economy and, thus, does not dominate the aggregate one. However, it should also be kept in mind that exchange-rate problems reflect high U.S. interest rates.

Taxes

Tax systems differ both in the amount of government services to be funded and in the choice of instruments, e.g., value-added, sales, and corporate and individual income taxes. When these differences are combined with international rules on rebating taxes on exports, the result may be to put a disadvantageous tax burden on one producer vis-à-vis a foreign producer. In particular, one can argue that the greater reliance on indirect taxes by Japanese and West Europeans has disadvantaged the Americans in the trade arena. The argument appears stronger for U.S./Europe comparisons than for U.S./Japan ones, though. In Japan, commodity taxes are a minor source of government revenue. Much of the disparity in the tax burdens between the production of the two auto industries may be due to the fact that the U.S. industry uses more labor per car—and at higher wages—than does the Japanese industry. The tax burden then becomes an additional aspect of the manufacturing cost differences.

Macroeconomic Issues: Demand

Energy Prices

The future of both the world order and the auto industry depends greatly on the future price of oil. Changes in oil prices affect the automobile industry not only through income effects on demand and changes in the costs of auto use, but also through shifting demand between sizes of vehicles. For the future, one can posit two plausible scenarios, each having different implications for the world economy and the auto industry. Neither promises a return to halcyon days, and the existence of widely varying possible futures is itself a problem.

On the principle that a "rising tide lifts all ships," the result of a continued erosion of oil prices would be positive for the world auto industry in general. It would result in a reduction of auto-industry unemployment through increased demand due to both enlarged real incomes and lowered costs of operating cars. This, in turn, would reduce the pressure for protectionism and ease industrial restructuring. It would also result in some shift in demand in North America back toward larger cars, in which the industry there is preeminent. This would not imply, however, a return to old patterns in the United States. First, there has been irreversible technical progress in auto design for fuel efficiency (see chapter 6). Second, government-mandated standards for fuel efficiency make unlikely the return of the very large cars of the past. Third, there is emerging Japanese competition in the, by present standards, midsize-car market.

A second scenario is for real oil prices to resume an upward trend. The consequences are obvious and stressful for the economies of the world and U.S. auto production in particular. The most important stress, however, is not in the scenarios themselves, but in the uncertainty about their relative likelihoods and the possibility that the world may shift rapidly back and forth between them, more rapidly than governments or firms can adjust their policies.

State of the Economy

Government macroeconomic policies aimed at a reduction in the rate of inflation, the composition of government expenditure, and the extent and distribution of taxes all, directly or indirectly, affected the absolute extent of auto demand in the United States. On balance, the near-term result has been a recession and a downward shift in the demand for automobiles. The presumably unintended consequence of apparently high real rates of interest in the United States has had supply effects through exchange rates and the cost of capital. A weak economy not only may have made adjustment necessary because of the resulting overcapacity but also may have made it more difficult. The accompanying deadweight losses are less when an economy is growing than when it is stagnating.

As the U.S. and the world economy recover, this will relieve some pressure for protection. Overcapacity will be reduced—though, as we have seen, not eliminated—as demand picks up, and the remaining necessary adjustment will be easier to accomplish. Even so, long-term growth prospects for the North American auto industry are modest, just as for the rest of the world. It is not clear when, if ever, the industry will again reach its 1978 peak sales of 15 million cars and light trucks.

Microeconomic Issues: Supply

Manufacturing Cost Differences

The technology of automobile production is sufficiently mature so that it is widely diffused throughout the world. As a result, national differences in manufacturing costs can play an important role in trade. In fact, the essence of trade is the arbitrage of such cost differences. Furthermore, as technology continues to diffuse, new participants who have even greater cost advantages may enter international trade. The only long-run solution is for the U.S. industry to resolve manufacturing cost differences. To the extent that it fails to do so, the probable outcome (in the absence of import restraints) will be that foreign producers will continue to increase their market share in those market segments where they continue to maintain an advantage. In these areas, this will force the domestic producers to accelerate offshore sourcing activities. If this outcome is unacceptable to policy makers, the alternative is permanent sheltering of the industry at great cost to consumers.

Microeconomic Issues: Demand

Perceived Quality Differences

Currently, automobiles imported to the United States are perceived by consumers to be of higher quality in terms of fit, finish, and reliability than are domestic cars, and available data supports this perception. In general, one can expect higher-income individuals to be especially sensitive to these factors because of the high cost of owners' time lost in attending to repairs. Also, one can expect the rate of increase in the cost of repair services to outstrip that in manufacturing.

Although the U.S. industry has made substantial improvements in its performance in these dimensions, continued progress is critical. If these efforts succeed, the perceived quality disadvantage will recede. There will, nevertheless, still be room for international specialization and trade in particular niches having to do with buyers' tastes.

Political Economy of Free Trade

The international liberal trade regime is fragile because of the dilemma that it is in each country's individual interest to violate the rules as long as its trading partners do not. If all countries defect, however, the open regime collapses and with it go the mutual gains from trade. It is because of this characteristic that the post-World War II order has been one of negotiated, reciprocal liberalization. Nevertheless, the liberal trade order may result in very concentrated adjustment costs as particular industries are disadvantaged. Because the costs of protection are diffused while the benefits are concentrated and because sharply focused interest groups have disproportionate political power, governments have a tendency to react, even if in the aggregate costs exceed benefits.

Since there are many possible aims of trade policy, individuals may advocate different or the same policies for different or the same reasons. The aim of policy may be to preserve existing returns to claimants on industry (shareholders, managers, and workers). Whatever the policy instrument chosen, the consequence is an income redistribution from consumers at large to these claimants. The aim of policy may be to change the terms of trade to the advantage (or lesser disadvantage) of the nation imposing the policy.

Again, the dilemma is that if enough important countries attempt this, trade will collapse. The aim of policy may be to preserve a manufacturing or technical capacity for national defense purposes. The case is arguable, but the argument needs to be made with reference to costs. The aim of policy may be to avoid excessive adjustment costs. That is, society may decide to forego the benefits of free trade if these are outweighed by the deadweight losses (as manifested by unemployment and lost taxes) that accompany the transfer of resources from the obsolete industry to other, more viable ones. It is important to remember, though, that the disappearance of an industry is not itself a cost unless it affects national security. Finally, the aim of policy may be to give an industry "breathing room" while it revitalizes itself. This is an intellectual variant of the "infant-industry" argument. To make this case for protection, one must demonstrate that the cause for the decline is surmountable.

Instruments

The policy instruments discussed below, tariffs, quotas, and local-content legislation, all represent what are in essence taxes on domestic consumers and subsidies to domestic producers. (Note that this does not mean that they cannot increase national welfare.) They differ primarily in their total cost, effectiveness, and political acceptability.

Because all policy instruments are *subsidies* in fact, the best, from the point of view of economic efficiency, is a straight cash subsidy. This tends to be politically undesirable precisely because the subsidy is then so visible. *Tariffs,* as taxes, are the next best instrument in that they permit market forces to continue to work to some degree. In particular, they encourage foreign producers to further lower their costs of production in order to maintain their exports. There are legal problems, though, having to do with most-favored-nation treatment under GATT.

The greatly increased efficiency costs of *quotas,* relative to tariffs, are well known. Quotas are inflexible. Long-term, they reduce competition and incentives to foreign producers to reduce costs and prices. They require arbitrary definitions of what constitutes a foreign product, and they may fail to provide the anticipated protection when the domestic market is in recession, unless they are in the form of a limitation of market share. Nevertheless, quotas are attractive to policy makers for several reasons. The key fact is that the quantitative effect, in terms of number of units, is certain, and they are better suited than tariffs for quick emergency application. Through the legal fiction of "voluntary export quotas," they bypass the GATT restrictions on the use of import quotas. Quotas are also attractive to policy makers because the importing country can persuade the exporting country to accept quotas by permitting it to retain the "monopoly profits" they create. Finally, because quotas are more effective at restricting imports than are tariffs, they are a better instrument for retaliation and bargaining.

Local-content legislation of the kind being proposed in the United States is a form of market-share quota (per car). If set at severe levels, these requirements ensure that the exporter will either abandon its foreign markets or locate inside the protected market. The benefits to U.S. auto workers and those dependent on the industry are evident. Local content, not incidentally, also limits the trade that would ensue from domestic producers sourcing components from abroad and thereby provides further benefits to those dependent on the U.S. auto industry.

Even if foreign producers export from the United States to gain local-content "credits" to offset additional imports, this merely represents a diversion of trade from its lowest cost source and is not trade creation. To the extent that the problems of the auto industry are related to high labor cost, the policy will tend to forestall adjustment. In fact, foreign firms that set up behind the barrier will come to support the policy's retention to protect their investment. One of the issues to be discussed in chapter 7 concerns the relative importance of high labor cost as a factor in U.S. industry problems.

For policy makers, local-content legislation is attractive in that it provides concentrated and visible benefits. It also preserves some competition that quotas vitiate. It is unattractive in that it imposes a high cost on consumers by preventing the arbitrage of costs, which is the essence of trade, and is destructive of the liberal trade order.

A major problem with any protectionist measure or, for that matter, a "managed trade regime" is that, once in place, it is hard to remove. Such cases can be found in both Japan and the United States. For instance, Japanese protectionist industrial policy in the 1960s—automobiles are a case in point—was liberalized far too slowly relative to Japan's world-class economic performance. Another problem is that the existence of protectionist measures may reduce the pressure to make necessary adjustments. There are many examples from around the world of import-substitution policies, which led to infant industries that never grew up. This, in turn, may depend on whether the industry becomes cartelized or not. In Japan, import substitution worked for a number of reasons. First, not only was MITI unable to impose consolidation of companies and market sharing on the auto industry, but there were too many auto firms that were able to maintain their independence. Second, there were company-wide "enterprise unions," not industry-wide unions. Thus, labor was unable to take full advantage of the protection to raise wages dramatically. Third, the government promoted exports, particularly through tax incentives, which forced manufacturers to meet world price and quality levels.

It is important, therefore, that if governments resort to protection, they choose instruments that encourage needed adjustments and that allow market mechanisms to work as much as possible in order to hold the costs of the policies to a minimum. Once adopted, any protection measures should contain an explicit schedule for their dismantlement in order to reduce uncertainty for the industry and to ensure that the industry accepts the inevitability of the need to adjust. The current "voluntary restraints" create particular problems because the uncertainty not only makes planning difficult but leads to continuous lobbying by the affected actors trying to change the outcome.

Short-Term Trade Politics

Since the "second oil shock" of 1979, the most explosive issue in the American automobile industry has been Japanese imports. This problem continues to threaten U.S.-Japan relations, although at the time of writing it had been kept within bounds by skillful handling on both sides of the Pacific. What follows here is a brief sketch of events and background factors.

Background

The postwar process of integration of the world economy, including Japan's rapid trade expansion and the decline of American economic primacy, has produced recurring trade "frictions" and even crises over an extended period. The typical pattern, seen in textiles, steel, color television, and other products, has been a sharp rise in Japanese imports coupled with economic problems in the domestic industry, leading to political demands for protection. The threat of formal legislation by Congress has led several presidents to pressure Japan into various export-restraint agreements. Specific trade issues have become linked in the public mind with other resentments against Japan—its perceived closed market, low defense expenditures, undervalued yen—to produce a generalized sense among many of unfairness and threat. Japan thus becomes a convenient scapegoat for American economic ills. The "textile wrangle" of the late 1960s demonstrates that even an economic problem of (objectively) relatively minor proportions can easily be mishandled to create major domestic political problems in both countries and a severe strain in the bilateral relationship.

Auto trade is by no means a minor economic problem. Capitalizing skillfully on a small-car market left unfilled by American manufacturers, and backed by the experience and scale economies acquired after a decade of rising domestic sales, the Japanese makers rapidly expanded their sales in North America during the 1970s. The rising market share of imports, increasingly dominated by Japan, brought sporadic calls for protection, but the healthy market for cars (over 9 million domestic cars sold in 1978) meant that these demands did not become serious until the Iranian crisis and subsequent shortages of gasoline in 1979. These events, plus the government's decision to decontrol gasoline prices, led to a sharp increase in demand for small, fuel-efficient cars. The cyclical economic downturn, hastened by governmental credit restrictions, brought a sharp reduction in total auto demand. The result was disaster for the U.S. auto industry: plunging sales and production, large losses in their domestic operations, the near bankruptcy of Chrysler, and—most important politically—soaring unemployment among workers directly or indirectly employed in the auto industry.

It would be astonishing if such conditions did not lead to intense demands for protection against import competition, and, indeed, during 1980 the United Auto Workers, some manufacturers, and many politicians from auto-producing areas were calling for help. The initial reaction of the Carter administration was to reject the idea of import controls—not, apparently, from a concern with the U.S.-Japan relationship, but because (based on an interagency economic analysis) it was concluded that protectionism would not produce many jobs and would harm the American economy by stimulating inflation and increasing the dependence on imported oil (since imported cars were more fuel-efficient). At this early stage, the political energy generated by the auto crisis was diverted into two other channels: a modest program of direct aid to the auto industry and its workers and pressure on Japan in areas other than export restraints.

This pressure concentrated on three areas, none of them new. First, the administration intensified its efforts to achieve relaxation of various Japanese safety and other standards that were held to discriminate unfairly against foreign auto exports. Second, various other actions aimed at increasing imports from the U.S., particularly of auto parts, were requested. Third, the Japanese were urged to build factories in the United States. The Japanese reactions to these demands, put forward in a series of negotiations in spring 1980, represented, as in similar past confrontations, modest accommodation. In particular, MITI endorsed the principle that Japanese firms with high exports should build factories abroad and exerted public and private pressure on the auto firms to invest in the United States, incurring some resentment. However, MITI had no power to force such investment decisions, and, except for the case of Honda (which had announced it would build a plant in Ohio), the officials were unable to overcome the worries of auto executives about the viability of manufacturing in the U.S. These negotiations, therefore, did not have a major impact in solving either the economic or political aspects of the auto-industry problem.

Pressure intensified. June 1980 saw both the passage of a symbolic Senate resolution stressing the increase of unemployment in the auto industry, cosponsored by eighty senators, and the filing by the UAW of an "escape clause" action with the International Trade Commission. Cracks within the administration were also appearing as Transportation Secretary Goldschmidt, who headed an interagency task force on autos, moved toward favoring import controls. And, of course, a presidential campaign was under way. Although both Carter and Reagan endorsed free trade, Japanese cars were too conspicuous to ignore, and both candidates were drawn into strong, albeit somewhat vague, promises of action.

Shortly after the election, the ITC decided by a narrow vote that Japanese car imports were not the primary cause of serious injury to the American industry. However, political realities made it impossible for the matter to end there. Pressures for action mounted quickly within Congress, and soon after his inauguration President Reagan dispatched Special Trade Representative Brock to Tokyo for discussions about providing relief. The result was the "unilateral" and "voluntary" export restraint announced by the Japanese government on 1 May 1981.

Perhaps the most noteworthy aspect of this process is that it proceeded so smoothly. In comparison with the textile case a decade earlier, the problems of the American auto industry were far more profound, and whether or not Japanese imports were the "primary cause," they were certainly enormous in volume. Nonetheless, despite some heated rhetoric, the conflict did not really get out of hand either domestically (in either country) or internationally. One reason was that neither the proponents of protection in the U.S. nor the opponents of restraint in Japan amounted to a unified and broad bloc. The positions of the American manufacturers varied with their market shares and their extent of overseas operations, and even the UAW took a more flexible position—deriving from both its "statesmanlike" tradition and its substantial membership in the export-oriented aircraft industry—than most labor unions would in like circumstances. Most officials in the Reagan administration believed in free trade even while they recognized the necessity for action on autos. On the Japanese side, MITI's orientation had shifted considerably over ten years to a more "internationalist" position, so it was less inclined to defend a domestic industry and its exports. Even the relatively sophisticated automakers (unlike the small, fragmented, but politically powerful textile manufacturers) were quite fearful of protectionist trends in America and around the world. Most important, although the various actors had conflicting interests and differences of opinion about the details of what should be done, it appears that by 1981 all realized that some action was inevitable. For that reason, the agreement could be worked out in a relatively cooperative fashion.

Even most of the form and content of the eventual resolution was not as controversial as one might expect. Most participants on both sides, including congressmen, wanted above all to avoid protectionist legislation such as a local-content law, which would create international legal and political complications as well as a snowballing of demands for relief from other American industries in trouble. The agreement had to be strong enough to defuse congressional pressures. Unilateral presidential action or a formal bilateral agreement would also have led to similar international complications from the perspective of both governments. A voluntary restraint, of the type already imposed in other product areas, was the obvious answer.

If the restraint was to be voluntary, it could only be a quota in terms of either market share or the number of units (a dollar ceiling would be impossible to enforce). After some debate, the American participants agreed to urge an absolute-number quota rather than a percentage share, based on what turned out to be an overly optimistic forecast of future demand. The quota, while voluntary from the American perspective, would have to be legally enforced in Japan to protect the manufacturers from antitrust suits by their own distributors in the U.S. In the end, only the number was really open to dispute, and number problems are the easiest to solve through a bargaining process—the final figure of 1.68 million units appeared to be reasonable in the light of American market projections at the time. Some arm-twisting by MITI was needed to obtain the agreement of the Japanese makers, and the ensuing battle over how the quota would be apportioned was quite intense. But again, both conflicts were substantially moderated by the general realization that something along those general lines had to be accomplished.

The voluntary restraint was quite successful in the sense that it resolved the bilateral issue, at least in the short run, and provided a tangible "accomplishment" for the administration that quieted American protectionist pressures for a time. However, its economic impact was considerably less than anticipated because of the American recession: auto sales were so poor that Japanese cars slightly increased their market share despite the quota. It should be noted, however, that price and demand data indicate that the Japanese did sell fewer cars than they would have otherwise, even in 1981 and 1982—although, of course, the quota framework meant that the resulting "excess profits" passed to the Japanese companies (rather than to the American government, as through a tariff). In addition, a continuing incentive was provided to upgrade imports in terms of both accessories and body size. In any case, the continuing problems of the American auto and other industries, particularly unemployment, soon led to renewed attacks on Japan and pressure for further solutions.

These surfaced in various arenas. In Michigan and other auto-producing areas, incidents of vandalism of Japanese cars and verbal and occasionally physical attacks on their drivers or on Asian people occurred. A few union leaders, businessmen, and politicians made extreme speeches that bordered on racism. While these relatively scattered incidents were overplayed by the press in Japan (where they were far more widely known than in the U.S.), there was clearly a heightened public concern, particularly in the industrial states.

Given the real problems and widespread public interest, it was inevitable that protectionist initiatives would appear in the legislative arena. The UAW and sympathetic congressmen urged local-content legislation that would force the Japanese to locate factories in the United States. Legislation of this sort passed the House of Representatives in fall 1982 and passed again in

1983, although in both cases it never came to a vote in the Senate. Moreover, complaints escalated about a variety of alleged unfair Japanese advantages, including a favorable exchange rate, a tax system biased toward exports, and "industrial-targetting" practices that had provided illegitimate government assistance; objections to nontariff barriers to the Japanese market also continued. Various pundits and politicians—including, particularly, Democratic presidential contenders—called for some sort of American "industrial policy" that might or might not include restrictions on imports, particularly including Japanese autos.

The response of the administration was to continue pressing for more access to the Japanese market but to resist local-content or other further protectionist policies with regard to automobiles, as well as "industrial-policy" interference with the free market. It did press successfully for a fourth year of the quota (at 1.85 million units) to defuse campaign pressures. The Japanese government could do little except to continue to dismantle trade barriers and to agree (beyond the terms of the voluntary-restraint agreement) to maintain the ceiling of 1.68 million units in the second and third years and then to accept a fourth year of restraints in late 1983.

Access to Japanese Auto Markets

As noted in the previous section, a great deal of the protectionist sentiment in the United States has been fueled by a sense of unfairness with regard to the access of U.S. automakers to the Japanese market. Currently, this discussion is largely symbolic rather than of practical importance. The issue had more substantive meaning in the past when the U.S. automakers were more of a competitive threat in the Japanese home market.

It is often forgotten in the U.S. public debate that the U.S. automakers have not had a significant direct export strategy to world markets in the postwar period. Rather, as noted earlier, they opted to export capital in the form of establishing production facilities abroad. This freed them from some of the charges commonly leveled against their Japanese counterparts with regard to dumping and exporting unemployment, but it also opened up U.S. business in general to charges of dominating the domestic economies of foreign countries. Indeed, this was the major fear of the Japanese throughout the 1960s and early 1970s.

In the context of the above remarks, the major barrier of the postwar period erected by the Japanese was not tariffs, commodity taxes, or nontariff barriers, but the prohibition on foreign equity investments in Japan. It was not until 1971 that, pressured by Western nations, the Japanese government agreed to drop some of the major barriers to establishing foreign-owned firms in Japan. Yet, Japan had achieved world-class status as an auto exporter as early as the mid-1960s. Since 1971, U.S. auto firms have advanced their equity shares of Japanese firms, with Ford holding 25 percent of Tōyō Kōgyō, Chrysler 15 percent of Mitsubishi, and GM 34.3 percent of Isuzu and 3.8 percent of Suzuki.

Generally speaking, the pattern in Japan has been one of gradually dismantling trade barriers in response to worldwide pressures. From the foreigners' perspective, this process has been painfully slow, with tariff levels having been kept at a high level far past when they could have been justified by the infant-industry argument. Similarly, a variety of nontariff barriers have frustrated Western manufacturers.

From the Japanese point of view, these changes have taken place with great rapidity in view of the tremendous fear that the Japanese have had of domination by large Western multi-nationals. A more subtle factor inhibiting trade liberalization on the part of the Japanese has been the "import-substitution" mentality of government bureaucrats and middle-management officials based on a 100-year-old ethic of catching up to the West. "Catching up" has meant the substitution of indigenous production for foreign exports and has been held up as the path to Japanese salvation.

The process of changing and reorienting the import-substitution mentality so that Japanese success worldwide will be seen as requiring the import of Western products is a slow and painful process. Whether Japanese progress in dismantling restraints on foreign imports has been fast or slow depends on one's perspective. From the point of view of unemployed U.S. workers and devastated midwestern communities, the progress is intolerably slow. From the point of view of Japanese history and the difficulty of reorienting national policy, it can be seen as remarkably fast. The facts, in any case, are that today Japan has no tariffs on imported foreign cars (contrary to the view of many "informed" American business leaders, not to speak of the general public). Tariffs stood at 40 percent on small cars in 1967 and fell to 20 percent in 1970, to 6.4 percent in 1972, and finally to zero in 1978. Auto parts followed a similar, though delayed, pattern.

The Japanese have also been criticized for a variety of practices such as the utilization of a commodity tax, which currently stands at 17.5 percent for small cars, and the absence of self-certification to meet safety standards. The United States has no commodity tax and does allow for self-certification to meet its safety standards. Thus, Americans criticize the Japanese for their lack of reciprocity. But it is also true that the Japanese practices are similar to those in Western Europe, where broad-based value-added taxes are imposed and self-certification is not practiced. Thus, from the Japanese point of view, they believe they are being unfairly singled out for what are common worldwide practices. There is no easy resolution to such conflicting perspectives. It is the case, however, that the specific enforcement of such rules and the procedures for setting rates, etc., have seemed to Westerners to discriminate against foreign exports. These latter issues can be addressed in an objective fashion.

It is often noted in the United States that the price of U.S. cars sold in Japan is commonly twice that of the U.S. retail price. This has been cited as clear evidence of Japanese obstructionism. Yet, it is clear that much of the price markup reflects the penalty for low volume. To be sure, one cannot dismiss the pattern of continuing petty harassment by bureaucrats from the Ministry of Transport; this has been reported as a common experience of foreign auto importers. The Japanese have announced a succession of liberalized import procedures, most recently in January 1983, designed to defuse Western criticism. From the viewpoint of foreign auto importers, the new procedures never seem to go far enough; however, significant movement has taken place since 1981. Japanese efforts include the dispatch of inspectors overseas as well as a shift in the objectives of the Japan External Trade Organization (JETRO). JETRO, whose prime objective has been to promote exports, is now expected to emphasize import promotion as well.

But such a focus on taxes and harassment misses the forest for the trees. As noted above, Americans have not had a strategy for exporting autos during the postwar period. Moreover, their strength lay in large cars, for which there was no mass market in Japan. Finally, the current competitive strength of the Japanese auto industry makes it clear that, with all conceivable import barriers removed, there would be no major increase in U.S. exports of auto products to Japan. It is for this reason that we suggested in our opening remarks that we are dealing with a largely symbolic issue. This is not to suggest that it should not be addressed. Symbols are powerful weapons in the formulation of trade policy, and a sense of unfairness exists as a festering sore poisoning Japanese-American bilateral relations. Moreover, it should be noted that export-oriented European automobile companies with presumably more suitable product offerings have had only modest success in penetrating Japanese auto markets. The Japanese public perceived these offerings as luxury products. It is difficult to know whether such sentiments could have been overcome by a more imaginative advertising strategy. In any case, it is clear that the various barriers throughout the postwar period have been quite real in their consequences.

A large imbalance in auto trade does exist between Japan and the United States. Relative to the strength of the Japanese auto industry, the Japanese government has been very slow in reducing their tariffs and eliminating other barriers on finished vehicles and auto parts. They have responded in a "foot-dragging" fashion, responding slowly to strong pressures, and only when imports were no longer a threat; this has inspired enormous mistrust on the part of their trading partners. There were, however, farsighted Japanese leaders who realized that it was in Japan's own long-term self-interest to liberalize capital and imports. Such leaders are said to have forcefully pushed these views and to have skillfully used Western pressure to achieve their ends.

It is important for the overall tenor of U.S.-Japanese bilateral relations that the last vestiges of trade restrictions in the auto area be eliminiated. This may require the Japanese to reexamine commercial practices that were not intended to restrict foreigners but nevertheless have had that effect. Given their enormous trade surplus in the auto area, they are under some moral obligation to "lean over backward" to insure that all remaining obstacles to Western auto imports be eliminated. Although the changes will not necessarily yield major substantive effects, they must be advanced with vigor so that the continuing sense of unfairness on the part of Westerners is eliminated.

At the same time, Western producers must recognize that, as import barriers have diminished, the initiative to export successfully to Japan is increasingly in their hands. To be successful, U.S. vehicle exports must meet Japanese consumer standards of quality, maintenance availability, and costs and be promoted by strong dealer organizations. Given the competitive issues cited above, however, we cannot expect the elimination of import barriers to have major effects.

Chapter 5
The Automobile Industry
and Public Policy

Introduction

Any large industry in any advanced nation is strongly affected by public policy at many levels. Macroeconomic and international trade policy help determine market demand. Monetary and tax policy and the management of the financial system influence the cost of capital. Antitrust policy structures relationships with competitors, suppliers, and customers. Educational and other programs affect the quality and availability of labor. Technological advances often rest on publicly sponsored research. Products and manufacturing processes are constrained by government regulations of many sorts. Some of these public policies apply almost evenly to all business, while others affect particular industries differentially; at the extreme, one finds "targetted" policies that are deliberately structured to promote or shape a narrowly defined industrial sector.

Our basic question is, How has the postwar development of the automobile industries in the United States and Japan been influenced by governmental actions? Our primary concern is with policies that differentially affect automobiles, whether positively or negatively, and whether deliberately intended or instituted for other reasons. The stereotypical view sees the United States as *laissez faire*, at least in orientation. Government interferes with markets as little as possible and especially avoids "industrial policies" favoring one sector over another. When government and industry do interact, the relationship tends to be distant, suspicious, and adversarial. Japan is characterized as a "developmental state" with an active governmental role in the market. Officials and businessmen work together to devise specific "industrial policies" that relate each sector to the development of the entire economy. The implication of these stereotypes is that the government's role in the development of the automobile industry has been both much more important and much more positive in Japan than in the United States.

Stereotypes are rarely without some basis in reality, but they rarely tell the whole story. A more accurate account begins with correcting some commonly held assumptions about the government-auto industry relationship in the two countries.

Misleading Assumptions

U.S.-Japan Similarities

First, the tendency to compare America and Japan without taking the rest of the world into account has led to an exaggerated perception of difference. Actually, there are important similarities. Unlike the European pattern, neither government participated directly in the management of an automotive firm: no nationalization, no direct investment, no "national champions." Also unlike Europe, the success of both industries was based on dominating their domestic markets when these were effectively insulated from foreign competition. In 1970 (and indeed still today), no other major country had so small a share of its market taken up by imports as Japan and the United States. For both industries, the main challenge of the postwar years has been "internationalization" in the sense of exposure to the world market—although obviously the form and results of this exposure have been quite different.

American Government Support

The second misleading assumption is that the American auto industry developed without much governmental assistance. Of course, the basic cause of its astonishing postwar growth was consumer demand, which was primarily the result of economic prosperity. However, per capita car ownership in the United States has always been much higher than that of any other country, even when disposable household income is held constant. Moreover, Americans drive larger, more powerful, and more expensive cars, which can be manufactured in quantity only in the United States. While America's wide open spaces and other natural factors largely explain these differences from the auto markets of other advanced nations, the fact that American public policies toward the automobile have also been quite distinctive is extremely important.

1. Fuel taxes at all levels of government have been far lower in the United States than in other advanced industrial nations throughout the postwar period. Moreover, for a set of reasons to which public policy has not been irrelevant, American energy costs, even exclusive of taxes, have historically been low.

2. The United States has invested enormously in building ever-better roads and highways for both intracity and intercity motor-vehicle travel. Both local and national governments have contributed. Investment in mass transit and long-distance railroads has been relatively lower than elsewhere in the postwar period.

3. The American financial system, which of course is heavily regulated by government, has functioned to supply large quantities of cheap consumer credit, making it easier for the average family to purchase cars. Credit costs are subsidized through the tax system.

4. Easy consumer credit and a variety of other policies at the local and national level have encouraged a pattern of dispersed suburban living in single-family homes, requiring one or more automobiles for normal life.

5. Low sales taxes and registration fees, the ease of obtaining a driver's license, lax regulations on auto inspections in most states, availability of parking, and so forth make owning and operating a car far more convenient and inexpensive in the United States than elsewhere.

In all these areas, government policy in Europe and especially Japan has been far less favorable to the expansion of auto demand.

Japanese Nontargetting

The third common assumption that needs correcting is that the Japanese auto industry was "targetted." There is no question that the Japanese government has targetted various industries for special promotion, attempting to guide and support them through financial aid, preferred access to foreign exchange, import protection, tax breaks, sponsored research, prevention of new entries, selective antitrust nonenforcement, and informal "administrative guidance."[16] In most cases, these industries were identified as critically important in national economic-development planning and were covered specifically by laws and formal ordinances. Detailed plans drawn up by appropriate sections of the Ministry of International Trade and Industry (MITI), in cooperation with the businessmen affected, had substantial impact on investment, pricing, R&D, production levels, competitive relationships, and market strategies in the targetted industries. A typical pattern has been the encouragement of rapid investment during economic booms, alternating with government-sponsored "recession cartels" to control prices and production when economic growth slowed.

Such targetted industrial policies were most important in the immediate postwar period, when they were applied to such "basic" or "axle" industries as coal, electric power, and steel. In somewhat more diffused form, they were later developed for petrochemicals and certain portions of the machinery industry; since the late 1960s the chief example has been computers and

---

[16]One may find a good analysis of industrial targetting in Jimmy Wheeler, Merit Janow, and Thomas Pepper, *Japanese Industrial Development Policies in the 1980's: Implications for U.S. Trade and Investment* (Croton-on-Hudson, NY: Hudson Institute, 1982).

related fields. Each of these industries plays its role in a developmental strategy of evolution from lower to higher value-added, from import substitution to exports, and toward higher and higher knowledge-intensive levels of technology.

It might well be supposed—and some later accounts by MITI and its enthusiasts have strengthened the impression—that the auto industry was designated to play a key role in this long-term restructuring strategy. Auto is, after all, a prime "linkage industry" that draws on a broad range of materials and technologies, and its growth—first domestically and then in the world market—is one of Japan's great industrial successes. However, a close examination of the historical record, supplemented by interviews with government and business leaders active at the time, reveals clearly that the automobile industry was not really "targetted." It was not specified in enacted industrial-development legislation nor mentioned prominently in economic plans. Its relationship with MITI has been much less intimate than those of targetted industries. Few support policies were specifically designed for autos, and of those, the largest were failures.

All this is not at all to say that public policy was irrelevant or trivial in the development of the Japanese auto industry.[17] The Japanese government has pursued many policies benefiting business, particularly manufacturing, and the relationship between government officials and businessmen is generally closer and more cooperative than in the United States. Our point here is simply that one must analyze the impact of governmental actions on nontargetted industries such as automobiles (or, to cite another prominent Japanese success, consumer electronics) in quite different terms than when attempting to explain the development of steel, aircraft, heavy chemicals, or computers.

Given that the Japanese and American auto industries are not as different as sometimes thought, that some public policies helped the auto industry in the United States, and that autos were not one of Japan's targetted industries, we may turn now to characterizing the real nature of the relationship between the government and the auto industry, beginning with Japan.

Government and Japanese Automobile Successes

The key period for understanding the development of the Japanese automobile industry is from the latter part of the 1950s until about 1970. Previously, in the decade following the war, little happened in the passenger-car field (domestic production, excluding the assembly of knockdown kits from Europe, rose to only 20,000 units in 1955). The automotive firms had

---

[17]Hiroya Ueno and Hiromichi Muto, "The Automobile Industry of Japan," *Japanese Economic Studies* 3.1 (1974): 3-89.

been eager to make cars and pleaded for government support, but the responses were grudging and *ad hoc* because Japanese officials were preoccupied with building up "basic industry." Much of the small amount of public-loan financing was less positive "industrial policy" than a means to secure foreign exchange or, in one case, to tide Nissan through a long strike. Japanese automotive companies established their strength in this period, but government had little to do with it—the key was the Korean War, directly through large "special procurement" orders for trucks by the U.S. Army, indirectly by providing a spurt to the entire Japanese economy.

In the succeeding period, auto production grew exponentially (domestic passenger-car production was 165,000 in 1960, 700,000 in 1965, over 3 million in 1970). Quality rose and costs fell to the point that the Japanese small car became fully competitive internationally. These developments must be carefully investigated if one is to discover the impact of governmental policy. Two approaches are possible. One is to examine all the tools of industrial policy, identify those that apply to automobiles, estimate the effect of each, and aggregate the results into a "net effect" of government. This method has advantages, but, as well as requiring large amounts of data, it risks overlooking the impact of other factors and thus exaggerating the importance of government—a not uncommon problem in many reports on Japanese industrial policy. Our approach is the opposite: we try to identify the key factors in the success of the auto industry in this period and then ask how these were related to governmental policy.

Our analysis suggests that three, possibly four, factors were crucially important: (1) The explosive growth of consumer demand. Production, company profits, and the level of investment are obviously most fundamentally related to how many people buy cars. (2) Protection from foreign competition. Without trade and investment barriers, foreign cars (probably European) would have sold extremely well, and large foreign auto firms (probably American) would have bought out Japanese firms or built their own factories. Production by domestic firms would have grown much more slowly. (3) Competition. The rivalries between Toyota and Nissan and among all the eleven firms that made passenger cars was intense through the entire period and, even without import competition, pushed each company toward higher quality, efficient production, and new and diverse models. (4) Though perhaps not quite so crucial as a cause of the auto industry's success, the concurrent and rapid development of auto suppliers, particularly the parts industry, contributed substantially to cutting costs and raising quality. We now ask, How were these factors related to public policy?

Explosive Consumer Demand

The Japanese government did not anticipate, much less plan, the motorization of Japan. For example, the highly expansive Income Doubling Plan of 1960—written when the upward trend of car sales had already begun—forecast that by 1970 Japan would have 21.9 cars per thousand people. The actual figure turned out to be 84.6 cars per thousand, *four times* higher. During this period and indeed today, Japanese national and local governments pursued the most stringent set of anticar policies found anywhere. Driver's license tests are hard to pass and require expensive lessons (now over $1,000). Parking is scarce, and in most of Japan a citizen must prove he owns a parking place to register a car. A rigorous safety inspection every two years costs upwards of $200 even for a car in good shape. Fuel is expensive, recently over $2.50 a gallon (including about 85 cents tax). Various taxes are high: an owner of a small (1600 cc) car would be charged 17.5 percent of the factory price (commodity tax) and 5 percent of the retail (auto-acquisition tax) at time of purchase, plus over $50 per year in weight tax and over $100 per year in auto tax (1982 figures). It was estimated in 1980 that the tax burden over ten years of ownership in Japan would amount to about $2,600, compared to about $450 in the United States.[18]

In general, Japan is the reverse case for the favorable conditions in the United States listed above. Highway investment has been substantial, but no one who has driven in Japan would argue that streets and highways are up to the American or European level, while intercity railroads and mass transit are exceptionally well-developed. Some more subtle effects are the direct or indirect result of a general bias against consumption and in favor of investment—for example, consumer credit was extremely scarce until recently, although the auto firms provided installment plans. All these policies have the net effect of holding down demand, certainly well below the level that would be reached if American-style policies had been implemented. In this sense, the explosive growth of auto sales occurred in spite of government policy and was due mainly to rapid economic growth and the strong desire of Japanese consumers to buy a car as soon as they had enough money.

Protectionism

That everyone bought *Japanese* cars—the second factor—was due to government policy. Tough trade barriers stemmed from an upsurge of imports (including purchases of used American cars from Occupation

---

[18]Shōgo Amagai, *Nihon Jidōsha Kōgyō no Shiteki Tenkai* [Historical evolution of the Japanese automobile industry] (Tokyo: Akishobō, 1982).

personnel) during the Korean War, draining scarce foreign exchange. In 1953 over 20,000 vehicles were imported; by 1955 the number was cut to under 5,000. Since Japanese firms could not meet the growing demand for passenger cars, licensing arrangements with Western firms to assemble knockdown kits were authorized, but foreign exchange controls were used to gradually increase the proportion of local content, up to 100 percent by 1960. Even at the outset, MITI's motives included "infant-industry" notions, but it should be noted that into the 1960s nearly all Japanese industries were protected. Foreign consumption goods were particularly shunned; the only imports welcomed were raw materials and machinery embodying technology that Japan did not yet possess.

Thus, although trade and capital protection was extremely important to the auto industry, for the most part it was less a "targetted" policy than a reflection of support for manufacturing in general. However, by the early 1960s, MITI officials were thinking much more seriously about automobiles and, along with industry leaders, worried obsessively about the threat from foreign cars and, particularly, foreign capital—as one former official put it in an interview, "The last thing we wanted was for the Japanese industry to become a subcontractor to the American car companies." When Japan moved more fully into the international economic community and lowered its trade barriers, automobile liberalization was delayed as late as possible (imports until 1965, and capital not scheduled until after 1970, although this stricture was broken—to MITI's consternation—by the Mitsubishi-Chrysler deal of 1969).

Competition

The main government response to the threat of internationalization was in the area of competition, our third factor. Here, too, public policy ran counter to actual developments. MITI officials have often seen virtue in economies of scale and evil in "overcompetition." In 1955, MITI's automobile section had announced a competition to design a small and inexpensive "people's car"; the winning company would be granted government support for production and, in effect, a semimonopoly in the largest segment of the domestic (and ultimately export) market. The officials were apparently influenced by the organization of Volkswagen in West Germany (*kokuminsha*, people's car, is a literal translation of *volkswagen*) and by a general concern that Japanese firms were too numerous and too small to compete internationally. Later, throughout the 1960s, MITI repeatedly tried to reorganize the auto industry into two or three large groups that would segment the market. Special loans were allocated as inducements to mergers, and the "Temporary Act for Specific Industry Promotion," aimed at expanding MITI control particularly over autos and steel, was introduced in the Diet. However, this bill died, largely because of opposition from the banks and other business interests. Despite strong "administrative guidance" from MITI, only a few auto tie-ups occurred; resistance was general within the industry.

We might note with some irony that although concentration ratios rose slightly from the mid-1960s, the auto industry in Japan remained highly competitive according to some measures (though not others), more so than the U.S. auto industry. Certainly, there is a greater number of domestic producers in Japan. This pattern prevails despite the fact that the U.S. government has consistently threatened and occasionally carried out anti-trust actions to counter concentration, while MITI did its best in the opposite direction in Japan—a good example of the limitations of public policy in both countries. In any case, the experience of the 1970s, when smaller Japanese auto firms like Honda, Tōyō Kōgyō (Mazda), and Fuji (Subaru) were highly successful in technical innovation and export marketing, indicates that MITI's fears about overcompetition may have been overdrawn; thus far, domestic competitiveness seems to have increased rather than diminished international competitiveness.

Targetted Auto Parts

The fourth factor, the development of supplier industries, has a less obvious but still important impact on the success of the auto industry and a somewhat problematical relationship to public policy. The steel case is the most clear cut: early development of the steel industry was a top government target, and inexpensive, high-quality Japanese steel has been an important input for automobiles (just as protected and expensive American steel has increased costs in the United States). Similar observations apply, to some extent, to other materials—aluminum, plastics, semiconductors, and perhaps ceramics.

The auto-parts industry was also targetted, though at a lower level of priority than the material suppliers. In 1956, it was one of seventeen fields specified in the "Law Concerning Provisional Measures for Development of the Machinery Industry," under which a government-business committee drew up three highly detailed five-year plans. The first plan listed production and cost-cutting targets for 75 separate parts. Japan Development Bank loans were provided to about fifteen parts firms every year ($50,000-$100,000 each during 1956-59), accelerated depreciation and tariff exemptions on new equipment were allowed, and administrative guidance encouraged the expansion of the scale of production and import of new technologies.

This quite specific industrial policy has generally been seen as successful. For example, half of the 25 percent cost reduction achieved by the auto industry from 1958 to 1960 is attributed to lower prices for parts. In addition, improvements in quality, reliability, and technical sophistication were notable. Table 3 is a measure of the impact of the program in its second stage:

Table 3
Impact of Development Bank Loans in the Auto-Parts Industry, 1960-1965
(units: average annual growth rates in percent; number of firms)

	Number of Employees	Capital-ization	Value of Production	N
Large Firms (capital at least $2.8 million)				
with loans	9.4	24.9	18.5	10
without loans	7.4	17.7	14.6	5
Medium Firms (capital $0.14 to 2.8 million)				
with loans	11.3	29.3	17.6	47
without loans	9.4	16.0	13.6	73

Source: Tsūshō Sangyō Kenkyūsha, *Nihon no Jidōsha Kōgyō,* 1965 edition.

The figures are annualized growth rates from 1960 to 1965, differentiating firms that received Development Bank loans and those that did not, among all members of the Japan Auto Parts Industrial Association.

Since few studies of Japanese industrial policy are this explicit about impact, these figures merit some consideration. At first glance, it seems apparent that the loans made a substantial difference. However, note that even the firms not receiving special aid grew at impressive rates. Moreover, if (as seems probable on the basis of other evidence) the Japan Development Bank decided which firms *within* the targetted industry should receive loans by taking normal banking criteria into account, it would follow that the stronger firms received the support. These would probably have grown somewhat more rapidly anyway.

More generally, as noted above, one should not analyze public policy in isolation from other possible factors. Two stand out. First is the market: the total value of motor-vehicle production was growing at a 17-18 percent rate in this period, a great incentive for expansion. Second, evidence from case studies and interviews indicate that the auto manufacturers were even more concerned than MITI with "rationalization" of their suppliers and that they were extremely active in providing technical and management support, some financing, and quite forceful advice. It is worth noting in this connection that another facet of MITI's policy was to encourage mergers among parts firms, leading to a pattern of large "specialized parts makers" producing narrow product lines but selling to several assembly firms. As with the ministry's attempts to concentrate the auto industry itself, this effort basically failed— the mergers that occurred were mostly within groups (some leading to product diversification), and the ties between assembly firms and their suppliers generally strengthened in this period.

On balance, it appears that, even within the explicitly targetted auto-parts industry, the impact of public policy should be seen as helpful rather than determining. Government loans and other aid were used effectively, but the industry probably would have developed along much the same lines even if MITI had not intervened.

Government and American Automobile Problems

Industrial policy is, thus, not the dominant explanation for the success of the Japanese auto industry. We now ask, To what extent do American governmental actions explain the problems of the American auto industry?

Insulation

Because of the American preference for large automobiles, for much of the postwar period the U.S. market was insulated from foreign competition. Exports were low because few outside the United States could buy large cars; imports were low because no foreign manufacturer could sell enough large cars at home to bring costs down to competitive levels.

This insulation turned out to have much the same effect as textbooks predict for formal protection. That the American auto industry had grown complacent and inefficient was well demonstrated by rising Japanese imports in the 1970s, by the revelations of the differential in landed costs in the early 1980s, and by the falling quality relative to Japanese competitors in several areas critical to consumer choice. Moreover, the high profits reported in early 1984, while due partly to the high prices allowed by the Japanese voluntary export restraint, were also reflections of the substantial drop in the "break-even" point brought about by drastic cost cutting. Clearly, up to the 1980s, considerable slack had developed in Detroit.

In retrospect, it appears inevitable that when a few giant firms compete only among themselves, the market will tend to stabilize and "mature." Product and process technology becomes standardized, wage rates are equalized across the industry, and competition—while seemingly intense—becomes focused on styling and marketing. Companies adjust their prices, with leadership from the largest, to cover costs plus an acceptable profit. Under such conditions, it is rational to sign generous contracts with the union rather than risk a long strike, and it is easy to allow bureaucratic management to proliferate and costs in general to creep up. Few saw much wrong with it until the market shifted dramatically to small cars and the Japanese demonstrated that cars of higher quality could be produced at much lower cost.

Why did protectionism of a more direct and comprehensive sort not have similar effects on Japanese carmakers? For one thing, Japanese domestic competition—as indicated by new entries in the market—was intense and was based on price and quality. A second reason, ironically, is that Japanese protection was by deliberate government action rather than apparently being "natural," as seemed true of American insulation. That artificial barriers can quickly be removed—and soon would have to be, given international pressures on Japan—was easier to predict than the likelihood of a world oil panic and its effect on the American taste in automobiles. Third, and perhaps most important, in the 1960s the American manufacturers "knew" they were the world leaders, while the Japanese "knew" they had to try hard to catch up. It was not until the "model" and the "challenge" effects had reversed direction that the American industry started on a new course of action.

Government officials did not play the main role in all this, but it is worth noting that MITI constantly harped on the threat of "internationalization" throughout the 1960s when no one in Washington was paying much attention to world auto trends. Still more to the point, as noted above, American public-policy choices, such as holding fuel taxes at the lowest level in the industrialized world, were not unimportant in insulating the American market from international competition. To that extent, the U.S. government shares responsibility for the softness of the auto industry.

Managing Internationalization

Inefficiency may be the most fundamental cause of the problems of the American auto industry, but the immediate cause was the substantial increase in fuel prices in the 1970s, which diminished insulation and exposed the domestic firms to international competition. This process was beyond the control of government, but government actions did affect its pace and its impact. The story is well known. Gasoline price hikes and, particularly, the transitional shortages that followed the 1973 OPEC export cutbacks created a furor among American drivers and propelled the energy problem to the top of the governmental agenda. However, solutions were not so obvious. The economists' preferred response was market pricing, but the prior existence of an oil price-control system meant that an explicit government decision to decontrol would be needed. Such a decision became politically more and more difficult amid growing public impressions that the shortage had been manipulated by giant oil companies to raise prices, and that further price hikes would be too big a burden on consumers, particularly low-income groups. The government accordingly maintained relatively low prices, but, motivated by a perception of long-term world petroleum shortages, a desire to escape dependence on OPEC, and political incentives to "do something," it also imposed the 1975 Energy Policy and Conservation Act. This regulation, which mandated progressive improvement in corporate average fuel economy (CAFE) over a ten-year period, created an artificial situation: the auto industry was forced to "downsize" without the market demand for small cars that higher fuel prices would induce. A glut of small cars resulted, and the industry was under intense cross-pressure when hit by the 1979 oil crisis and the subsequent economic recession.

Looking back at the 1970s, it is clear that neither the industry nor the government was prepared for the internationalization of the domestic market. The gradually though erratically rising market share of imports since the 1960s had been seen as a pesky but basically insignificant trend in Detroit and was almost completely ignored in Washington. Governmental and corporate decisions in the 1970s showed little consideration for the issue of international competitiveness. Here, the contrast with Japan is stark, where, from the late 1950s, the threat of internationalization was prominently mentioned in all discussion of the automobile industry among businessmen and government officials alike.

In any case, the post-1979 automobile crisis did bring relatively quick responses from the American government, notably, the Chrysler rescue, effective pressure for the Japanese voluntary export restraint, and a modest relaxation of environmental and safety regulations. Without arguing the merits of these specific actions, we observe that they represented no fundamental shift in American industrial policy. Regulatory relief simply softened earlier government interventions. The loan guarantee for Chrysler was regarded as a temporary exception, legitimated by the earlier Lockheed and New York City cases. Trade protection was the traditional response to an industry in trouble: it had many precedents, and the mechanisms were already in place. Although the more fundamental problems of the industry were becoming apparent—a plausible list might include high labor costs and low productivity growth, conflictive industrial relations, inefficient manufacturing practices, insufficient innovation, and the prospect of a severe shortage of capital—there were few serious suggestions from either government or industry about how government might help deal with them more directly. Nor was very much done to soften the impact of the sharp drop in auto sales on the people and cities most affected, although a few programs for job retraining and regional economic redevelopment were initiated.

The terms of the Chrysler rescue are worth noting, however. The mandated agreements to reduce costs by the firm itself, the United Auto Workers, and parts suppliers somewhat resemble Japanese government policies toward industries in trouble for either cyclical or structural reasons, particularly the latter. These, too, have included loans and some temporary trade protection, as well as production limitations to maintain prices and fairly substantial programs for job retraining and regional revitalization. The key is that explicit commitments for restructuring on an industry-wide basis are required. Such "positive adjustment policies" are difficult to implement and recently have come under criticism in Japan, but many in the United States have begun to call for a somewhat more systematic approach to industrial problems than *ad hoc* responses to the crises of the moment.

Other Governmental Actions

American public policies have, of course, impacted on the auto industry in many additional ways. Some have been supportive: as well as the actions noted above that supported high demand, these might include such programs as assistance from states and localities in building new factories; government-sponsored research in robotics that has disproportionately benefited automotive manufacturing; and unemployment insurance plus trade-adjustment assistance, which maintained the auto work force during layoffs caused by the boom-or-bust pattern of auto sales. There have also been various negative impacts; of these, the most prominent are the environmental, safety, and energy regulations of the 1960s and 1970s.

The legitimacy of antipollution regulations is not seriously contested: auto emissions as well as stationary-source industrial pollution are classic externalities, imposing costs on citizens that cannot be compensated or prevented through market mechanisms. Because it is in the interest of no single firm or consumer to bear the cost of emission-control equipment, government action is justified. There is, of course, room for considerable debate about whether a particular pollutant really causes much harm, whether a particular regulatory method is the best available, or whether the costs of a given level of regulation—the "last 5 percent" problem—are worth the benefits. Experts have argued convincingly that better planning and design of American environmental regulations could have produced better results with fewer burdens, but most evaluate the overall impact positively.

Safety regulations are on shakier ground conceptually since it can be argued that the costs of accidents are born primarily by the purchaser of the automobile, and he or she should be left to decide whether extra safety equipment is worth buying or using. Regulations to limit property damage (the five-mph bumper) are particularly controversial. However, proponents argue that accidents are externalities, in that they do impose costs on society, and that, as a practical matter, building safety into all automobiles is quite inexpensive compared with the benefits.

Fuel-economy regulations have neither the solid theoretical foundation of antipollution standards nor the practical arguments behind safety rules. Certainly, holding gasoline prices below market prices had no justification; if conservation beyond that enforced by the market is taken as a policy goal, either raising prices through taxation or direct rationing makes more sense than forcing manufacturers to defy trends in consumer demand. This fundamental contradiction in public policy emerged briefly once before but became moot with the 1979 oil crisis; at time of writing it has reemerged as a controversial and extremely difficult issue.

Some economists and politicians are proposing the removal of the CAFE regulation before GM and Ford incur the penalties (financial and public relations) for exceeding the mileage standard. If CAFE is abolished, as has been advocated in Detroit and Washington, and if fuel prices do not rise much, it is likely that the trend toward larger cars will continue. In effect, the American market would be partially reinsulated by its unusually low fuel costs. American automakers clearly would benefit in a market weighted toward the segment with the least import competition, and where profits are higher. A logical corporate strategy for the U.S. domestic market would be to devote attention and resources to this segment, competing with Japanese small cars mainly through captive imports or the products of joint ventures.

The obvious danger in this scenario is that, if the American industry once again commits itself to large cars, it is extremely vulnerable to another oil panic. However, difficult problems for future public policy might well occur even without such a major shock. It is questionable whether this strategy could work without substantial long-term trade protection, since in a free market the Japanese cost advantage would allow aggressive pricing of intermediate cars, gradually eating away at the market share of domestic producers. Indefinitely maintaining trade barriers, in turn, would have serious implications for both the U.S.-Japan relationship and for the world liberal trading order.

These are possible future problems. What impact have these environmental, safety, and energy regulations had on the American auto industry to date? The costs of research, retooling, and equipment attached to cars have been substantial; moreover, dealing with regulations has absorbed time and energy that might have been devoted to meeting the Japanese competition. However, three qualifications are needed. First, by far the most expensive adjustment was "downsizing" to meet energy regulations, but much of that presumably would have been needed to respond to the market even if the government had taken no action. Second, technology-forcing regulations probably led to research with unexpected benefits, such as computer control of engine function, originally aimed at emissions but important for economy and performance as well. More generally, Detroit engineers have noted that their status and the attention given to technical innovation within the auto firms rose markedly with the need to meet new regulations. Third, the safety and environmental regulations had little influence on competition between American and Japanese cars because, of course, they applied to both and because Japan's domestic emissions standards are at least as stringent as those in the United States.

In evaluating the extent to which governmental actions have contributed to the current problems of the American automobile industry, we find most important, first, the policies reinforcing earlier market insulation, which left the industry vulnerable, and, second, the inept handling of the energy crisis, which aggravated rather than alleviated the difficult transition of internationalization. The impact of environmental and safety regulations appears a lesser order of magnitude. Even adding up the individual negative influences of each of these policies, we would not conclude that government should be assigned more blame than either market forces or auto-industry management. However, we are impressed that the American auto companies were faced with a great many pressures all at the same time in the past decade, including ups and downs in total demand, gyrations of consumer preference, new competition from imports, high interest rates, and capital shortages, along with a host of regulations (including pollution and safety standards for factories as well as products), rapid shifts in government pronouncements, and barrages of public criticism. This turbulent environment, partly the result of American government actions, certainly hampered management attempts to deal with the industry's fundamental problems.

Conclusions

This chapter has focused on the impact of governmental policy on the auto industries of Japan and the United States. For the most part we have discussed "industrial policy" in the sense of actions with a particular and differential effect on the automobile industry (whether intended as such or not). Our approach, as much as possible, has been to weigh these policy factors against the various other economic and managerial factors that seem most important as explanations for the course of development in the two industries. Given current concerns, it is natural that this has chiefly meant trying to explain Japanese successes and American problems. Perhaps the most valid finding to be drawn from our study is that the impacts of these policies have been varied, very complicated, and occasionally perverse. It is also fair to conclude, however, that in general we do not see industrial policy as quite so decisive an influence as has sometimes been argued, either positively in the Japanese case or negatively in the American case.

It is important to emphasize the somewhat narrow scope of our discussion. Because of the current interest in specific or "targetted" industrial policies, and because of resource limitations, we concentrated on governmental actions rather than systems and did not attempt to assess how much difference public policy in a broader sense has made on industrial development in the two countries. It is quite reasonable to believe that such factors as macroeconomic management, the structure of the court system, legal rules governing corporate behavior, the operation of financial institutions, the biases of tax systems, quality and quantity in public education, labor law, international economic policy, antitrust laws and their enforcement, and so forth have been extremely important to the two auto industries.

However, if an auto expert were asked to list the factors that currently give the Japanese car industry its competitive edge, he or she might well mention the following: lower wages; better organization of the work force; employee commitment; thoroughgoing attention to quality; close relations to auto suppliers; more automation; elimination of inventories through the "just-in-time" system; emphasis on both product and process innovation; close cooperation between sales, manufacturing, and research; skillful market development, particularly overseas; flexible manufacturing; and high and rising productivity in general. Without further discussion, we observe that *none* of these factors was primarily caused by the Japanese government, and few of them even have very much indirect relationship to public policy.

Finally, with regard to the relationship between government and business, we observe MITI's function of monitoring problems in the auto industry and relating these to broader trends in the Japanese economy. Whether or not the officials' proposed solutions were the best, or even were implemented, may be less important than the maintenance of this broad and future-oriented perspective—in our interviews, even auto-industry executives who generally downplayed the impact of concrete governmental actions remembered well MITI's exhortations about internationalization in the 1960s. No public or private body played this role consistently and effectively in the United States. A related point is that, because of Japan's geography and history, the question of international competitiveness has always been much on the minds of officials and businessmen alike. A real lesson from Japan, then, is that more discussion, from a longer-term perspective, involving government officials along with businessmen and specifically addressing the position of the American auto industry in the world, might substantially improve its capacity to respond to future change.

Chapter 6
Market Factors Influencing the
Automobile Industry

Introduction

The future of the global automobile industry depends on the underlying economic conditions that influence automobile ownership. Although much of our attention in this study focuses on recent product and manufacturing innovations, the long-term outlook for the automotive industry is conditioned by the fact that it is a maturing industry. Worldwide, the growth of the automobile population decreased from 7 percent per year in the early 1960s to 4.5 percent per year in the late 1970s. This trend may well continue. For the advanced industrial nations in particular, this will make the demand for replacement vehicles an increasingly important part of the market. This, in turn, implies that new-car sales will become more susceptible to the business cycle and that the longevity of cars will have a greater influence over new-car sales. A replacement market will also imply that product refinement will play a greater role because of the sophistication and experience of buyers.

In comparing the U.S. and Japanese car markets, it is important to stress the different roles that automobiles play in providing transportation services in the two countries. For example, the automobile plays less of a role in moving people in Japan than in the U.S., and trucks and water transportation play a greater role in moving cargo in Japan. These differences stem in large part from the differences in geography and population densities between the two countries. Another important factor is the comparatively cheaper energy of the United States, and its greater domestic supply.

World Auto Demand and Economic Growth

The most fundamental fact regarding automobile ownership is that it is largely determined by income. This holds for particular countries over time, as the experience of both Japan and the U.S. has shown. More dramatically, it also holds for comparisons across countries. Table 4 presents the data on automobile ownership and per capita income in 1980 for selected countries. Auto ownership varies by a factor of over 200. The U.S., Canada, and Australia have roughly one car or truck for every 2 people, while countries with the lowest per capita incomes have one car for every 100 to 400 people. Clearly, the relationship between vehicle density and income is not perfect. In

Table 4
Automobile Ownership and Per Capita Income

Country	Population per Vehicle in 1980	Per Capita Income in 1980	Predicted Annual Growth in GNP, 1982-1995
U.S.A.	1.5	$9,521	3.6%
Canada	1.8	10,296	4.0
Australia	2.0	7,720	3.1
W. Germany	2.5	9,278	2.7
France	2.5	8,980	2.7
Italy	3.0	6.914	3.2
Japan	3.1	8,460	3.9
U.K.	3.2	7,216	2.5
Spain	4.2	5,500	4.1
Argentina	6.5	2,331	4.2
Czechoslovakia	6.6	3,985	3.0
Greece	7.3	4,590	3.8
Portugal	8.2	2,000	4.7
Brazil	12.0	1,523	5.2
Mexico	15.0	1,800	5.6
Malaysia	19.0	714	5.2
Taiwan	42.0	1,300	5.2
Philippines	46.0	779	4.7
Egypt	76.0	448	5.7
Indonesia	127.0	415	4.4
Pakistan	225.0	280	4.4
India	432.0	150	3.9

Sources: *MVMA Motor Vehicle Facts and Figures, 1982, The World Almanac and Book of Facts, 1983*, and *Worldcasts* (published by Predicasts, Inc., 20 May 1983).

part, this represents differences in population density and government policy, as well as problems involved in correctly measuring income. Table 4 also provides predicted annual rates of economic growth for 1982 to 1995. These tend to be lower for the high-income countries. An important implication of the data in table 4 is that increases in income have a greater impact on less-developed countries. For example, a doubling of income from $1,500 per year to $3,000 per year might be expected to create a threefold increase in vehicle ownership. On the other hand, a further doubling of income from $3,000 to $6,000 would probably result in only a doubling of car ownership. Consequently, the greatest percentage growth should be expected in low-income countries, especially since they have higher expected rates of economic growth.

For countries with 1980 per capita incomes in excess of $6,000, it seems reasonable to expect growth rates in vehicle ownership that are roughly one-half of their expected growth in income, roughly 1.5 percent per year (see Working Paper 14 for assumptions). In contrast, low-income countries should experience growth in vehicle use greater than their income growth, perhaps in the range of 7 to 10 percent per year. It is useful to keep in mind, however, that this growth is relative to a small base. The United States alone accounts for 38 percent of the world's motor vehicles, Europe, including eastern Europe, accounts for another 36 percent, and Japan 9 percent. In contrast, the rest of Asia and Africa, which have 66 percent of the world's population, account for only 6 percent of the world's vehicles. Based on this analysis, it is estimated that worldwide auto registrations will grow at less than 4 percent per year.

Replacement Markets

The number of new cars bought each year can be broken down into two categories: those bought to replace cars that will be scrapped and those that represent net additions to the fleet of all cars. Net additions to the stock of cars, expressed as a percent of the total fleet, have been in the range of 1 to 3 percent over the last decade in the United States and have fallen steadily since 1950. In contrast, Japanese net additions in the 1970s have been in the range of 4 to 12 percent per year, but have fallen from over 40 percent in the early 1960s. In other words, the Japanese market has matured much faster, and this maturation will continue to proceed at a rapid pace. If the experience of other countries of similar population densities is a guide, one vehicle (including passenger cars and trucks) for every 2.5 people would be a likely saturation point, the point at which growth in the fleet of vehicles slows to keep pace with population growth. In 1982, the ratio stood at one vehicle for every 2.85 people. At current rates of growth, roughly 5 percent per year, the 2.5 saturation point would occur by 1985. However, the Japanese growth rate may well continue to decline, implying saturation at a later date, perhaps in ten to fifteen years. Still another possibility, reflecting the distinctive expansion of the minicar market in Japan, is that the total number of vehicles in the market would not level off until a stock of 60 million is reached. Assuming the same 5 percent growth rate, the 60 million total would be reached around 1990 and would leave Japan with a saturation level of around one vehicle for every 2.02 persons. At the point that saturation is reached, the replacement market would play at least as large a role in Japan as in the United States.

At first glance, market saturation would appear to foster market stability since the key remaining factor is population growth. The driving-age population is predictable years ahead. However, replacement markets are inherently more volatile. In countries with rapid rates of economic growth and low levels of automobile use, the net additions to the stock of automobiles occur in response to long-run increases in the standard of living; the business cycle plays comparatively little influence. However, in advanced countries with comparatively little economic growth, the replacement market is heavily influenced by short-term business cycles. The number of registered cars does not change much, but the mix between new and used cars is fairly volatile. Although owning an automobile becomes something of a necessity, owning a new automobile is still a luxury. This volatility of replacement markets has important implications for manufacturers because of the cost of changing rates of production and of carrying spare capacity to meet peak demands. It also implies that drops in demand, such as those experienced during the period 1979-82, should not automatically be interpreted as representing a fundamental, long-term shift in auto-buying habits and tastes. Rather, a severe recession coupled with the continued evolution of the United States to a replacement market caused the decline in new registrations.

Energy and Transportation

The two dramatic increases in oil prices of the 1970s have had the effect of focusing considerable attention on the possibility of still higher energy prices. By historical standards, oil prices (corrected for inflation) are not very high today. The real price of gasoline in 1977 was lower than in 1957, and gasoline prices in 1983 were roughly at the level they were in 1930. Nevertheless, the possibility exists that there might be another increase, in part because of political instability in the Middle East. However, even taking that possibility into account, we should expect the real price of oil to increase with time because of the change in the value of money over time. Those who hold oil should be indifferent between selling it today and investing their money, on the one hand, and holding on to the oil and selling it later, on the other. This sort of reasoning leads to the view that energy prices will rise if those selling oil are acting in their self-interest. In practice, of course, the actual price can fall because of unexpected decreases in demand, new discoveries of oil, or, in the case of a cartel, price cutting and dissolution. Factors that might increase prices faster than expected include increases in demand, wars, or cartel action. Still, although the potential range of prices is great, the best single guess (a probability weighted average, say) is that oil prices will tend to rise over the long term.

In looking at energy prices, there is a natural reason for focusing on petroleum and its derivatives. Since oil is inexpensive to transport (per Btu), oil prices are determined by global conditions. Consequently, the prices of other sources of energy that are more difficult to transport are determined by the world price of crude oil, the nature of local supplies and requirements, and transportation costs. Since these other factors are fairly stable, changes in the price of oil have a major influence on other energy prices. The mechanisms that provide this relationship are the shifts in consumer demand that occur when relative prices change and, more importantly, the shifts among energy sources, particularly for stationary users such as power plants. The full significance of this must be taken into account. If the only use for oil were to fuel automobiles, and the only use for coal were to generate electricity, the primary mechanism by which increases in oil would be reflected in coal prices is through the increase in demand for electricity that could occur as people spend less on gasoline. These effects are not likely to be large. In practice, the more forceful effect will come about because certain applications can readily be converted from one use to another in response to small variations in price. This means that as the world price of oil rises, the whole cluster of energy prices rise along with it, with variations across fuels depending on the ease of substitution.

This picture of the energy market would appear to explain why coal-based liquid fuels have not yet posed a serious threat to gasoline. It is true enough that, if oil prices rise and coal prices stay the same, a point will be reached where it becomes economically feasible to use, say, coal-based methanol instead of gasoline. In practice, however, relative energy prices are unlikely to change much because of the many other applications in which substitution away from petroleum is easier than for automobiles.

Energy Use in the United States and Japan

The data on current energy consumption by sector give a rough idea of the relative ease with which energy from different sources can be shifted to different uses. Table 5 breaks down total U.S. energy consumption in 1978 into four categories: residential and commercial, industrial, transportation, and electricity generation. (Electricity, of course, appears again in the last column as an energy input for the other three sectors.) There are some

Table 5

U.S. Energy Consumption by Sector and by Primary Source, 1978

(in trillions of Btu)

	Total	Coal	Natural Gas	Petroleum	Nuclear and Geothermal	Electricity
Final Consuming Sectors	78,014	3,849	16,490	33,881	—	23,760
Residential and Commercial	29,296	277	7,678	7,145	—	14,197
Industrial	28,129	3,572	8,274	6,737	—	9,511
Transportation	20,589	—	538	19,999	—	52
Electric Utilities	23,725	10,371	3,293	3,905	3,045	X

Source: *Statistical Abstract of the United States, 1979*, p. 602.

Note: The total for final consuming sectors does not include electric utilities since this sector's output appears as a separate source under electricity.

distortions in this pattern of use because of the U.S. energy policies in effect in 1978. The major policies were price controls on oil and gas. In the case of controls on oil prices, the chief effect was to discourage domestic production and encourage offsetting imports. The net effect on oil use was probably small. So, although American oil imports rose from 41 percent of total oil consumption in 1973 to 52 percent in 1978, despite higher world oil prices, domestic production fell over this period. The effect of price controls on natural gas was also to restrict production, but since it is more costly to transport than oil, the policy resulted in rationing, "shortages," and inefficient allocation for many years in the residential and industrial sectors. This policy is the major source of bias in these figures and probably results in a total for natural gas that is low compared to the other sources and, perhaps, a bias (on a percentage basis) toward uses in sectors with price controls since users there who had established "rights" to cheap gas had no incentive to conserve or switch to other sources of energy.

Despite these drawbacks, table 5 does provide some insight into what the likely effects of a rise in the price of crude oil might be. One would naturally expect a decrease in the percentage of total energy derived from petroleum, but this would vary by sector. The greatest decreases are likely to appear in electricity generation and industrial uses, where coal and gas already have a sizeable application. It might also be possible to expand residential and commercial uses of coal, perhaps by utilizing it in pulverized form. With more detailed data on the costs of various applications, more precise predictions of a response to increases in the price of crude oil would be possible.

As shown in table 6, the primary energy demand in Japan in 1979, expressed in oil-equivalents, was 432 million kl (kiloliters), of which 71.5 percent was supplied by imported oil. Eighty-three percent of this amount was consumed by industry and the personal sector, 14 percent by domestic transportation, and 3 percent by international transportation. Table 7 represents the breakdown between the industrial and personal uses. There is a slight difference between table 6 and table 7 owing to the different data sources, but it is apparent that consumption by industry accounts for a large share, roughly 40 percent of the total. Compared with the United States, where energy consumption is fairly equally distributed among industry, personal use, and transportation, the relative energy consumption by industry is much greater. The oil consumed by transportation corresponds to 29 percent of the total oil consumption. Automobiles alone consume only 20 percent of the oil in Japan, a quite different situation from that in the U.S., where 53 percent of the oil is consumed by automobiles. The important difference, however, is that the Japanese import 99.7 percent, effectively 100 percent, of the oil they consume. The absolute amount has been decreasing since the sudden rise in the oil price in 1979, which has been the result of the efforts made by every sector to conserve energy. The differences in energy situations are apparent in the fact that annual oil consumption in the United States is 2.6 tons per car per year and in Japan, 1.3 tons per car per year. The reasons for such a difference are easily found in the smaller and more fuel-efficient Japanese cars, as well as in the lower yearly mileage per car.

A few remarks are in order concerning electricity, natural gas, alcohol, and other sources of energy. Japan depends on oil for generating 60 percent of its electricity. Therefore, if it changes its present oil-consuming cars to electric cars, the total energy conserved would be very small compared to the United States, where the ratio of oil-dependent electricity is much lower. Natural gas, if its sales system is well established, can be widely used, as, even now, LPG is used by most taxis in Japan. (This results largely from efforts to reduce pollution, however, and it is by no means clear that it can be cost effective for the entire fleet of vehicles in Japan.) As for alcohol, the technology for its application to internal-combustion engines has been developed, and there are very few technical problems standing in the way of its general utilization. However, Japan depends on imports for 100 percent of its energy, whether it is oil, LPG, or an alternative energy such as alcohol. Quite unlike the United States, Japan has to be better prepared to respond immediately to external conditions such as the availability of imported energy and changing energy prices.

Table 6
Primary Japanese Energy Supply in 1979

Total Primary Energy Demand	432 million Kl	
Kind of Energy	Amount	Percentage
Hydraulic Power (ordinary) in 10,000Kw	1,879	5.2
Hydraulic Power (pumping-up)	952	
Geothermal Energy in 10,000Kw	13	0.0
Domestic Oil & Natural Gas in 10,000Kl	304	0.7
Domestic Coal in 10,000t	1,779	2.8
Atomic Energy in 10,000Kw	1,513	4.2
LNG in 10,000t	1,457	4.8
Overseas Coal in 10,000t (ordinary overseas coal in 10,000t)	5,939	10.6
New Energy & Others in 10,000 Kl	48	0.1
Subtotal in 100 million Kl	1.23	28.5
Imported Oil in 100 million Kl (LPG in 10,000t)	3.09 (977)	71.5
Grand Total in 100 million Kl	4.32	100

Source: Japanese Society of Automotive Engineers, *Newsletter* 36.1 (1982).

Table 7
Japanese Energy Demand by Sector in 1979

Sector	10^{10} Kcal	Percent
Primary Energy Demand	381,772	100.0
Reconversion	110,878	28.0
Final Energy Demand	270,894	71.1
Industry	149,962	39.3
Domestic Life	55,940	14.7
Transportation	57,924	15.2
Nonenergy Sector*	7,079	1.9

Source: Japanese Society of Automotive Engineers, *Newsletter* 36.1 (1982).

*Refers to the use primarily of petroleum to produce synthetic products such as plastics and synthetic fibers.

Fuel Price and the Mix of Cars

The leading influence in the United States on the choice of vehicle type has been the price of fuel. This is clearly evident in figure 7, which presents the real inflation-adjusted price of gasoline for the years 1955 to 1979 and the market share of imports and small domestic cars. The sharp price rise in late 1973 was followed by a considerable increase in the market share of both small domestic cars and imports. A similar increase in fuel prices in 1979 again produced an increase in the market share of these two classes of cars. The influence of energy prices is also evident in the late 1950s and early 1960s, when the market share of both imports and the newly introduced domestic compacts, such as the Corvair and Falcon, were high. (Shares for the latter group are not shown for the early 1960s.) Coincidental with the move to smaller cars, standard-size cars decreased their share from 36.8 percent in 1970 to 10.2 percent in 1981. The market share of U.S.-made compact and subcompact cars have moved together and are combined in figure 7. Their share peaked in 1974, dropped in 1975 to 1977, and has recently risen again. The market share achieved by U.S. luxury cars has increased steadily, but slowly, from 2.9 percent to 4.9 percent, but standard-sized cars have declined

FIGURE 7

The Real Price of Gasoline
and Vehicle Type

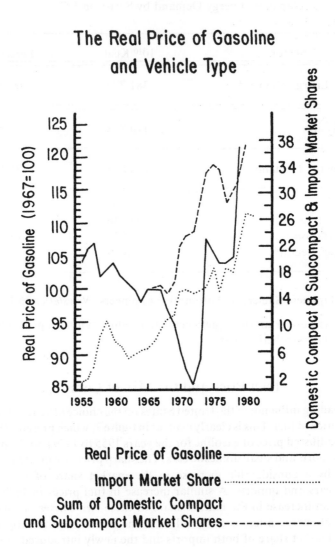

Real Price of Gasoline ————

Import Market Share ·······················

Sum of Domestic Compact
and Subcompact Market Shares – – – – – – –

SOURCES: *The Economic Report of the President*, January
1980, p. 261; M. B. Fuller, *Note on the World Auto Industry in Transition*, Exhibit II; and *Automotive News*,
1981 Market Data Book Issue, 29 April 1981, p. 18.

steadily through the 1970s (except for 1976 and 1977), from 36.8 percent in 1970 to 10.2 percent in 1981.

Although the association between fuel prices and the type of car that is purchased may be strong at times, it is also significant that, for a period beginning in the mid-1960s, the market share of both small cars and imports rose despite falling fuel prices. There are at least two plausible reasons for this: falling prices for small cars and foreign cars (adjusting for their attributes) and changes in demography and income that made increases in less-expensive smaller cars the natural response to (a) rising family incomes (because more families chose to own a second car) and (b) increases in the number of young people who found smaller cars more attractive. So, while fuel prices are part of the story, they are not the complete explanation. Moreover, future increases in the price of oil will not cause similar shifts in buying patterns. The case abroad illustrates that given increases in energy prices are becoming less important. European and Japanese gasoline prices were already fairly high relative to the United States in the early 1970s, and buying habits have not changed as much as in the United States. This can probably be traced to the increasing inconvenience and hazard associated with smaller automobiles and less-powerful engines. The implication of this is that future increases in world oil prices are unlikely to result in dramatic changes in the mix of automobiles desired by U.S. consumers. As miles per gallon increase and other benefits, such as room, ride, safety, etc., are lost, there is a trade-off that begins to limit consumer willingness to accept greater efficiency. At $1.50 per gallon, for example, a car that gets 10 mpg and travels 10,000 miles per year would use 1,000 gallons of gasoline at a cost of $1,500. At this level of fuel price and usage, if such a car were traded for one that got 20 mpg, the price of fuel would drop to $750 per year, a substantial and significant savings. However, a change to 30 mpg would result in an additional savings of only $250, and a change to 40 mpg would add only $125. Obviously, if these incremental improvements in miles per gallon were associated with an impractically small or uncomfortable vehicle, there would be little reason to make the change. Even at $2.00 per gallon, there is an incentive of only $266 per year to move from a car getting 30 mpg to one getting 50 mpg.

Because of the declining cost benefits from fuel economy and the recent decline in real fuel prices, there has been considerable concern that American customers may abandon small cars in favor of gas guzzlers. This is unlikely. Between the 1973 and 1982 model years, the actual miles per gallon of new cars increased from 14.2 mpg to 27.0 mpg, an increase of 90 percent. The consumer's contribution to this increase is considered to lie in his or her willingness to buy smaller cars. The manufacturer's contribution lies in the manufacturer's ability to improve the fuel economy of cars in other ways than reducing their size. If, in 1982, consumers had bought the same mix of sizes that they bought in 1973, they nevertheless would have realized an improvement in fuel economy from 14.2 to 24.4 mpg because of manufacturer improvements in the vehicles. Put another way, of the 12.8 mpg improvement between 1973 and 1982, 2.6 mpg was due to the consumer's choice of smaller cars—20 percent of the total—and 10.2 mpg—80 percent of the total—was due to manufacturer improvements. If, because of cheaper fuel, consumer demand created a size mix similar to that of, for example, 1978, the gas efficiency they would realize would drop to 26.4 mpg, a decline of 0.6 mpg, a loss of 2.2 percent. If consumers went back to the 1972 mix, their miles per gallon would still be 24.4, a decline of 2.6 mpg or 9.6 percent. As a practical matter, because of limits on product availability, consumers cannot go back to the 1973 mix.

The Role of Automobile Prices in Determining Import Shares

A central issue has been to what extent the increased market share of Japanese cars in the United States and other parts of the world is due to changes in the relative price of Japanese cars, and to what extent they are due to technological innovations. Clearly, consumers are willing to trade technology for price. The important policy question is the degree of substitutability. One way to approach this question is to look at the pattern of imports in the United States. States with a large fraction of imports tend to be those with comparatively low transportation costs for Japanese automobiles, such as Hawaii (64 percent in 1980) and California (52 percent), or those with high gasoline costs, such as Connecticut and Vermont (both 36 percent). Similarly, low import shares occur where transport costs for U.S.-made cars are low, mainly the Midwest, or where gasoline is comparatively cheap, as in Louisiana and Arkansas (both 20 percent in 1980). Marketing efforts are also important, although it is not clear to what extent marketing depends on cost factors. Similarly, geographical considerations also appear relevant, with large American cars being well-suited to driving conditions of the midwestern terrain. Notwithstanding the impact of such factors, there is limited evidence that variations in the delivered price of Japanese cars cause substantial changes in purchases, but the exact degree of this association is open to question.

Similar conclusions emerge from the pattern of imports over time. There is some evidence that the increased market share of Japanese cars has been associated with movements in the exchange rate and changes in relative prices, but a good deal of the changes cannot be associated with price movements. For example, the large increases in market share of the late 1960s were accompanied by increases in the relative price of imported cars. On the other hand, the relatively strong dollar of the last several years has probably contributed to increased imports from Japan.

Conclusion

The world market for automobiles overall is increasingly mature, although there will be significant areas of growth in selected developing countries. Increases in auto use will be below those seen in earlier decades. This is a general development and can be seen in both the United States and Japan. The maturation of the Japanese market, in particular, will occur rapidly because of the nature of Japanese economic growth and the inherent limits in the density of vehicle use. Japan and the United States also face different problems in transportation because of the disparities in energy supplies. Both, however, will have a more volatile, technologically demanding auto market.

Chapter 7
Manufacturing Cost Differences

There is an elusive but highly publicized number that has come, for much of the public and industry, to symbolize the competition between the North American and Japanese automotive industries. That number is the difference in manufacturing costs, the extra dollars required to build an automobile in North America compared to what it costs to build a similar vehicle in Japan. The various analyses that provide estimates of the MCD, as we shall call it, have crystallized the North American industry's concerns about Japanese competition, and the factors identified as sources of the MCD have either guided, or at least reinforced, U.S. industry efforts to address the cost disadvantage.

Concern about the possible implications of the MCD has led the U.S. industry to reexamine its standard operating procedures. The growing concern about efficiency in manufacturing processes, product quality and reliability, and improved labor utilization and relations arises from the potential competitive implications of the MCD, as well as the already impressive expansion of the share of the U.S. market by Japanese manufacturers. As reflected in lowered break-even points, the U.S. auto producers have made substantial improvements in raising productivity over the last few years. There is also evidence of improved quality levels, and these various efforts to raise productivity and improve quality can be expected to continue.

However, for the U.S. industry, the Japanese market share over the last few years has been more of an immediate problem than the MCD, since the Japanese have not yet sought to increase market share by drawing upon the MCD to reduce prices. With an automobile quota in effect since 1981, of course, they have had little incentive to do so. They have been able to sell all the automobiles allowed under the quota. From the American producers' perspective, their weakened competitiveness in North America vis-à-vis the Japanese has denied profits to U.S. manufacturers from a significant sector of the depressed markets of the past three years. This has forced U.S. manufacturers to hold employment at lower levels and to go to expensive (by historical standards) capital markets to fund their own product and manufacturing programs. They have also had to use large-car profits to subsidize their unprofitable small-car operations.

If the Japanese were to utilize the MCD in a price-competitive fashion, the continued existence of the U.S. automotive industry in its current form would be highly problematic. This has led the industry to pursue a variety of strategies, including possible restructuring in the form of decreased vertical integration through more outsourcing of parts and components. Much of this outsourcing may involve nondomestic manufacturers in order to take advantage of lower costs. Moreover, without automobile quotas, it would most certainly include an increase of "captive" imports of finished vehicles.

Since current U.S. prices of equivalent U.S. and Japanese vehicles are comparable, the existence of a large MCD suggests higher unit profits for the Japanese industry. This might yield capital for the Japanese industry to undertake technological, design, and capacity alterations that might convert to numerous further competitive edges in areas such as quality, reliability, and styling. These retained earnings could also be used to finance the penetration of Japanese vehicles into other markets than North America through price-competitive strategies. This would provide an edge in opening new markets and make North American penetration of these markets more difficult, whether through export or the more traditional strategy of the North American industry to locate production facilities abroad.

This chapter will review a series of reports that provide estimates of and identify sources of the MCD. We elected to pursue this objective rather than develop our own estimate of the MCD for a variety of reasons. The most important factor was the large number of these reports and the wide circulation given to their results; this suggested the utility of critically reviewing and evaluating them. The details of company-produced estimates have not been available, so the detailed review is largely restricted to public reports. While the review is critical as to both the consistency of these reports and the clarity with which they highlight sources of the MCD, it is not supportive of the argument that the MCD is either nonexistent or too small to be of competitive importance. Alternative estimates of a particularly critical source of the MCD—labor-content cost—developed from a variety of Japanese sources are presented. Finally, the Japanese compensation system is described, and its implications for cost levels and other competitive advantages is discussed.

Analytic Problems

Both the development of an overall MCD estimate and its allocation to specific sources reflect particular analytic decisions made by the authors. The proper comparison of manufacturing costs for different companies is not an easy task, and a number of assumptions and procedures necessarily underlie any particular set of estimates. First, vertical integration—the proportion of the vehicle actually manufactured by the producers—must be equated. To compare the costs of GM producing 60 percent of a vehicle with Toyota's production of 30 percent would not be especially useful. Second, differences in manufacturer's product mix of large and small vehicles must be eliminated since it is possible that the manufacturing costs are inherently different in the two cases. Third, capacity utilization must be equated, since the unit production costs, and many of its components, from capital to labor cost, vary with the proportion of plant capacity in use. All of these differences, if not properly eliminated or controlled, will bias the comparison in the direction of a larger MCD.

Once these factors are accounted for, and useful data secured, there are further problems facing the analyst with regard to the attribution of components of the MCD to specific sources. This is particularly a problem with the allocation of labor-content cost differences to productivity differences, on the one hand, and to wage-rate differentials, on the other.

A final problem confronting analysis of manufacturing cost differences is the determination of the proper exchange rate to use for developing the comparisons. The rather volatile movement of the yen-dollar exchange rate suggests caution in the use of any particular MCD estimate. The problem this raises is that the fluctuation in the exchange rate operates upon the estimated costs of the Japanese manufacturer, and not directly upon the difference between U.S. and Japanese costs. The actual impact of changes in the value of the yen upon the MCD, then, depends on the amount of change and the ratio of Japanese costs to the MCD. Where the MCD is large compared to Japanese base costs, changes in the yen-dollar exchange rate will have smaller effects upon the MCD; where the MCD is small compared to Japanese base costs, currency fluctuations will have larger impacts.

The problems of capacity utilization, vertical integration, product mix, cost allocation, and exchange rates are not simply problems for the analyst. Once the analyst has introduced appropriate procedures for addressing these problems, the decision maker still needs to assess the implications of the analyst's choices. That is, the decision maker must incorporate these factors and the reported analytic results into scenarios of likely futures.

Estimates of the Total Cost Difference

The reported MCD between the U.S. automotive industry and the motor-vehicle industry of Japan are presented in table 8. All these reports appear to be based on an exchange rate of between ¥210 and ¥220 to the dollar. The article by Professors Abernathy, Clark, and Kantrow (ACK)[19] actually presents three different estimates, derived by different methods and relying

---

[19]W. J. Abernathy, K. B. Clark, and A. M. Kantrow, "The New Industrial Competition," *Harvard Business Review* (September-October 1981), pp. 68-81.

Table 8
Reported Manufacturing Cost Differentials
(in dollars)

Source	Estimated Cost Differential
Abernathy, Clark, Kantrow (ACK)	
Table A (total)	1673.00
Table B	1973.00
Table C	1304.00
Abernathy, Harbour, Henn (AHH)	2050.00
Chrysler	2200.00
Ford (F)	1800.00
General Motors (GM)	1500.00
Harbour—Subcompact & Compact (HSC)	
Studied portion	983.00
Estimated total	2163.00
Harbour Revised (AHH)	1861.00
Harbour (DOT)	
Studied portion	1186.00
Estimated total	2488.58
Mateyka (M)	2109.00
Harbour (1983)	2203.00

upon different data sources and assumptions. Similarly, Harbour's (HSC)[20] study of the production cost differences in the manufacture of compact and subcompact vehicles provides two estimates. One covers the portion of the production process actually reviewed, and the other a straightline estimate from the studied portion to the full manufacturing process. The Harbour (DOT)[21] estimate is compiled from a series of four separate analyses of constituent manufacturing processes and their summary. A final process estimate from Harbour (HSC) was added to these to allow an extension to the entire OEM process, as in Harbour (HSC). The paper by Abernathy, Harbour, and Henn (AHH)[22] presents its own estimate, as well as an estimate attributed to Harbour that appears to be a revised estimate from HSC. Chrysler[23] and Mateyka[24] present only one estimate each.

In addition to these central papers, we have added estimates from three other sources. The estimate labelled "Ford" represents a figure frequently cited by Ford Motor Company representatives in public discussion.[25] The estimate labelled "General Motors" reflects the figure presented by GM in a recent publication (1982),[26] although recent public statements by GM officials estimate that it is either less than this[27] or as much as $2,000.[28] The Harbour (1983) estimate is drawn from a recent journalistic presentation.[29]

If we drop the two lowest estimates, which may reflect a more restricted portion of the manufacturing process than do the others, the full range of the

---

[20]J. E. Harbour, "Comparison and Analysis of Automotive Manufacturing Productivity in the Japanese and North American Automotive Industry for the Manufacture of Subcompact and Compact Cars" (Harbour and Associates, Inc., 1980, 1981).

[21]J. E. Harbour, series of four reports to the Department of Transportation, n.d.

[22]W. J. Abernathy, J. E. Harbour, and I. M. Henn, "Productivity and Comparative Cost Advantages: Some Estimates for Major Automotive Producers," Harvard Business School working paper (1981).

[23]Chrysler Corporation, "U.S. Tax Policy Gives Japan an Unfair Break," n.d.

[24]J. A. Mateyka, "Productivity: The Automotive Industry Worldwide," Paper presented at Annual Meeting, Society of Automotive Engineers, Detroit, Michigan (23 February 1982).

[25]For example, Jack Barnes, "Meeting the Japanese Auto Challenge: An American Progress Report," Paper delivered to Manufacturing Council, American Management Association, Tucson, Arizona (March 1982).

[26]General Motors Corporation, Public Interest Report (1982).

[27]Interview with Roger Smith, *Automotive News* (13 December 1982), p. E-6.

[28]John F. Smith, Jr., "Prospects and Consequences of American-Japanese Company Cooperation," in Robert E. Cole, ed., *Automobiles and the Future: Competition, Cooperation, and Change*, Michigan Papers in Japanese Studies No. 10 (Ann Arbor, MI: University of Michigan, Center for Japanese Studies, 1983), pp. 19-26.

[29]John Holusha, "Why G.M. Needs Toyota," *The New York Times* (16 February 1983), pp. 27, 29.

estimates is $1,300 to $2,200. This is not a small range, but one that, at first glance, might reasonably be accounted for by differences among data sources, variables, and analytic strategies pursued by the various authors. Yet they are so often repeated that they are taken as independent confirmations of the size of the cost difference rather than as a range of estimates. They have come to have, for both the industry and the public, an aura of precision, accuracy, and reality beyond what they can reasonably support.

This is not to say that there is no difference in manufacturing costs between the U.S. and Japanese industries. There is, and it is substantial. At the same time, the low and high estimates have different implications for both the magnitude and urgency of the responses required by U.S. manufacturers.

The reports from the Big Three reflect estimates based on their own analysis of proprietary data. The Project has not had full access to either their data or methods—only the Chrysler estimate provides much information about exactly how the calculations were made—so little can be said about them. However, the discussion of the published estimates should raise questions whether or not these reports provide independent confirmation of the proprietary studies. Many companies that lack the resources or access to information that is necessary to conduct studies of their own specific competitive situation may rely on these published reports as indications of both the magnitude and sources of the cost differential they might face. There are some problems with these studies and some potential limitations to the generality of their results.

Reports that address the costs of the Japanese to actually land a vehicle in North America use a range of $400 to $500 for this cost. The Japanese advantage in landed costs would net these costs out; it remains a substantial source of potential competitive advantage.

Detailed evaluation of these reports, where possible, are presented in the Joint U.S.-Japan Automotive Study Working Paper 3. Only a summary of major points can be presented here.

The three estimates provided by ACK in Tables A, B, and C of their report (see table 8) have received wide circulation in the industry and among the general public. All three cover only the cost of labor content; other sources of U.S. advantage or disadvantage are omitted. Table A presents an estimate uncorrected for vertical integration or capacity utilization, the lowest estimate of differentials in productivity, and the second lowest estimate of the ratio of Japanese to U.S. wages. No Japanese data as such is utilized, and the analysis reflects the estimated U.S. costs and their distribution, as well as comparative assumptions about productivity and wage rates. Table B is uncorrected for vertical integration, product mix, or capacity utilization and, therefore, represents a starting point for analysis, not a useful MCD estimate. Table C focuses on two companies only, ignores capacity utilization, and adjusts the estimates for product mix and vertical integration. These adjustments, however, are at variance with the adjustments for these factors in AHH (as are the basic data) although they are reported to be from the same source.

Harbour's HSC estimates are probably corrected for capacity utilization (as in Harbour DOT), and an examination of common products and processes introduces some level of control for product mix and vertical integration. He goes substantially beyond labor content and incorporates a wide range of cost factors in his analysis. However, in developing his estimate of the total MCD, he ignores differences in vertical integration and winds up either doubly counting a portion of differences associated with wage rates or simply overestimates it by a factor of two.

Abernathy, Harbour, and Henn ignore capacity utilization and adjust costs to a 50 percent level of vertical integration and to the Japanese small-car product mix. The order in which the data are adjusted, however, reintroduces a higher level of vertical integration for the U.S. producers and, thus, about 11 percent higher costs. More seriously, it results in an overestimate of the cost of labor content by about 30 percent and an underestimate of the differences in material costs by nearly 50 percent compared to the proper adjustments.

The estimate of Harbour incorporated in AHH is meant to reinforce the landed-cost estimate of AHH. However, the closeness of the estimates primarily reflects the cross-inclusion of estimates and a double count of wages in the Harbour estimate. If we eliminate those, AHH's estimate is some 74 percent higher than Harbour's.

The Chrysler paper raises a major issue in trying to develop a clear, consistent picture of the MCD. This is the high degree of variability in the factors different analysts do or do not take into account, as well as variability in analytic procedures. It is doubtful that one can safely add different factors from different analyses to arrive at a grand total. Hence, GM's and Chrysler's estimates of the cost difference, $1,500 and $2,200, respectively, might be quite compatible if we add the $650 tax disadvantage from the Chrysler analysis to the GM estimate. The problem is that GM, in all probability, has included at least some of these tax differences (for example, employer taxes on wages) in its cost estimates. One must be careful, then, in assuming that similar numbers constitute independent verification, and one must be cautious about taking cost estimates for specific factors from different reports and adding them together to arrive at a grand total.

The estimate provided by Mateyka is yet another rework of the Harbour HSC estimate as revised by AHH. It suffers from a double counting of wages and, in fact, expands it to the point that the Japanese worker must make negative income for the estimates to be internally consistent. There are other errors in the compilation, including errors of assignment to categories and the inclusion of U.S. advantages as disadvantages.

In summary then, AHH, HSC, Harbour revised (AHH), and Mateyka all produce MCD estimates larger than the underlying data support. In the case of AHH, this is only about 11 percent; it is substantially higher in the others, reaching 38 percent in Mateyka.

On balance, these studies all show a significant MCD, even if we correct for procedural errors. The MCD is probably closer to the $1,500 level, however, than to the $2,000 level produced by some of these studies. It is also important to note that these studies are based on data from 1979-81. Even though the Japanese industry presents a moving target and there has been adverse movement of the exchange rate, it is difficult to believe that the efforts of North American manufacturers resulting in lowered production costs, coupled with the altered contracts with the UAW, have had no impact on the size of the MCD.

Factors Involved in Estimating Costs

Introduction

These papers cover a wide range of factors that may contribute to the overall advantage of Japanese motor vehicles in landed costs. Technological factors, both process and product, receive attention, as do various aspects of the cost of the physical plant, including size, land cost, construction cost, maintenance, and energy costs. Differences in quality-assurance philosophy and practices between the two industries are examined for the implications on the work force of different inspection procedures, the costs of rework and scrap, and the warranty costs associated with differences in product quality.

The time and cost implications of different modes of labor utilization are explored: relief practices, absenteeism, layoff policies, work rules and associated task specification and responsibility, and approaches and structures to worker participation (such as Quality Control Circles)—all are given analytic attention. Differences in the role of inventory are considered, and estimates of cost differences associated with carrying charges, materials handling, and maintenance requirements are presented. Cost differences associated with purchasing materials, components, and tools and equipment are evaluated. Wage-rate differentials are considered, as is the elusive, overall factor of "productivity." Transportation costs are estimated, including those associated with supplies, components, subassemblies, and shipment of finished vehicles. Taxes and tariffs also receive attention.

A major problem with the estimates displayed in table 8 is that it is difficult to determine the extent to which the variation among them simply reflects consideration of different possible sources. Hence, it is difficult to assess the extent to which they are either compatible or inconsistent. Mateyka, for example, drawing on Harbour's earlier work, estimates a differential in landed costs of $1,709, remarkably close to the revised Harbour estimate (1983) of $1,718. However, we find differences between these two totals in the factors considered, as well as major differences in cost estimates within categories. Mateyka attributes about $770 to better management systems, while Harbour's later work attributes almost twice as much to that category. Mateyka assigns $812 to wage and fringe-benefit differences, while Harbour (1983) reports $550 for this category, in spite of an adverse movement in the exchange rate. Mateyka attributes $430 of the total cost differential to differences in material cost, but this category is completely absent from Harbour's later estimate. In other words, although there is agreement between the total estimates, substantial disagreement exists in the estimates for individual categories.

Fluctuations in currency exchange rates present a particular problem for allocating the MCD to different factors. For some factors, such as labor-content costs, the reported differences are very large compared to the Japanese base cost. For others, such as material cost, the reported differences are very small compared to the Japanese base cost. ACK's Table C, for example, presents an MCD based only upon labor cost, which is 220 percent of the Japanese base cost. AHH presents a material-cost component of the MCD for which the difference is 20 percent of the Japanese base cost. Movement of the yen from 240 to 180 per dollar (a 33 percent increase in value) would lower the ACK estimate of the difference in labor costs 13.1 percent, but it would lower the AHH difference in material costs 92.8 percent, with all other factors remaining constant. Thus, currency fluctuation has less of an impact upon differences in labor costs than differences in material costs. Because of this, the proportional allocation of the MCD to labor cost and material cost will alter as the exchange rate fluctuates.

Productivity and Wage Rates

There are two factors that are included in virtually every estimate of the difference in manufacturing or landed costs: labor hours required to produce a vehicle and the wage costs associated with those hours. For each of these reports, "labor content" accounts for over 50 percent of the total MCD. As is clear from the GM and Chrysler example given above, these estimates reflect both the way in which researchers allocate various costs and the factors that they include in deriving an estimate of the total. Although other factors, such as material costs, may be critical, they are not covered in all analyses. But all the reports identify the specific cost differences of "hourly productivity" and "wage." An examination of these categories will clearly delineate two disadvantages that decision makers must address and will illustrate some of the problems with these analyses of the differences in manufacturing costs.

The ratio of Japanese to U.S. productivity in the automobile industry, which underlies the cost estimates displayed in table 8, range from about

1.20:1 to 2.40:1; that is, the Japanese industry is estimated to produce anywhere from six to twelve vehicles in the labor time it takes the U.S. industry to produce five. The cost difference associated with these hourly productivity differences ranges from about $170 to about $1,060. In these reports, the percentage of the total difference in manufacturing costs accounted for by hourly productivity ranges from about 10 percent to about 54 percent. Both these ranges are quite large and certainly suggest less consistency in these studies than has been assumed. In these studies, the cost difference attributed to wage rates ranges from about $400 to about $1,300, about 25 to 80 percent of the total cost difference. Again, wide variation rather than consistency is the picture that emerges. If the cost difference associated with both productivity and wage differences is analyzed, consideration must be given to the difference between the total production hours multiplied by the wage rate for the U.S. industry and the total production hours multiplied by the wage rate for the Japanese industry.

The most useful and defensible way to allocate these total costs to productivity and wage rates is to recognize three components: pure productivity—U.S. excess hours charged at Japanese labor rates; pure wage—Japanese hours charged at the difference between the wage rates; and a joint effect—the difference in wages during excess U.S. hours of production. To assign the joint effect to either productivity or wages or to split the difference between them, as some of the reports do, obscures rather than clarifies the role of these two factors (see Working Paper 19). The important point, of course, is that a reduction in the difference in either productivity or wage rates will also bring savings associated with the joint effect. This is important when evaluating strategies for cost reduction, especially when it is likely that rapid changes in the industry will occur.

Table 9 displays the cost differences associated with labor content and the allocation of this cost to a productivity, wage-rate, and shared effect. The proportions in parentheses reflect the allocation reported or implied in the original reports. The other proportions reflect the allocations using the same three-way allocation. Even using a common allocation method and eliminating differences in the total cost of labor content, there still is a range of about 3:1 in the estimates of the relative significance of each factor. As was the case with total estimates of MCD, the reports are less consistent than they are typically treated. Here, in particular, they offer quite variable prescriptions for focusing ameliorative actions upon either wage rates or productivity deficits.

Productivity. A few comments must be made about the general issue of productivity comparisons between the Japanese and U.S. industries, both about the focus of these productivity comparisons and their generality. These reports by and large calculate labor productivity in terms of unit output per hour of work. Some attention is paid to nonlabor costs associated with this

Table 9
Proportional Allocation of Total Productivity/Wage
Cost Difference to Productivity, Wages,
and Joint Effect

PAPER	Cost difference due to productivity/wage factors	Proportion due to productivity (b-a)c	Proportion due to wage (d-c)a	Proportion due to joint wage/ productivity (b-a)(d-c)
ACK - Table A (OEM only)	$ 856.00	.13 (.20)	.74 (.80)	.13
ACK - Table B	1973.00	.35 (.54)	.27 (.46)	.38
ACK - Table C	1304.00	.25 (.38)	.48 (.62)	.27
AHH	1910.00	.37 (.50)	.35 (.50)	.28
AHH (Rev. 1)	1437.00	.29	.48	.23
Chrysler	1352.00	.46 (.52)	.23 (.48)	.31
GM	1500.00	.18 (.18)	.60 (.60)	.22 (.22)
H (DOT-Studied)	740.00	.48 (.76)	.24 (.24)	.27

measure, and some attention is given to unit costs other than labor. But the thrust of most of these analyses is to measure productivity and associated cost differences between the two industries by the differences in labor costs for the time required to produce an automobile.

Much of the effort to improve productivity in the Japanese and U.S. automotive industries has involved the substitution of capital equipment, such as transfer lines and robots, for human labor. That means that unit labor productivity is only a partial productivity measure. It is subject to over-estimating the difference in production costs when labor productivity is increased through other mechanisms that are themselves costly (whether more or less costly than labor only affects the size of the overestimate, not whether this measure will produce an overestimate).

Another aspect of this measure that needs to be kept in mind is that it does not take into account the differences in value added, which may be a direct result of differences in labor productivity. If a customer is willing to pay more for a more labor-intensive process (hand-sewn versus machine-sewn seat covers, for example), then the unit productivity of the workers in the two different processes becomes a much less meaningful comparison.

Finally, this measure of productivity is affected by many factors. That does not necessarily make it a bad measure, but it does suggest that we need to be cautious about what it does and does not tell us. For a variety of reasons, an industry running at or close to capacity will exhibit higher labor productivity than one that is operating well below capacity. Because of this, we need to be careful that our comparisons of labor productivity are telling us what we think they are. Comparative labor productivity measured at a particular point in time may tell us less about the fixed structural differences between the two industries than we think it does, and more about the general economic conditions of the two industries than we are aware it does.

These U.S. reports estimate, on the average, that the Japanese industry produces about 1.9 units in the labor time it takes the U.S. industry to produce one unit. This would drop a bit due to the slight decline of production and the stable work force in Japan over the last two years. But even if these estimates are accurate for the specific comparisons made, questions about their generality can certainly be raised. Time-series data from the Japan Productivity Center's 1982 report suggests that since 1979—the year for which most of these U.S. estimates were made—the unit productivity of the Japanese industry has been about 140 percent that of the U.S. industry. Another time-series extrapolation from Japanese data puts the Japanese level lower—about 10 percent above the U.S. industry. While these estimates are quite different, they are both considerably below the estimates derived for most of these U.S. reports on the difference in manufacturing costs.

One possible reason for this discrepany is that the U.S. estimates focus upon the OEMs and tend to focus upon the relatively more efficient Japanese assemblers, such as Toyota, while the Japanese estimates are industry wide. It may be that differences in labor productivity between the industries depend upon whether we examine assemblers or suppliers. While some OEMs have concluded that the Japanese supplier industry contributes to the MCD (see chapter 9), a recent report suggests that the productivity differences at the supplier level are less than at the OEM level.[30]

The Japanese tend to emphasize two alternative measures of productivity—value added per hour worked and value added in terms of labor and capital. Using a value-added definition of labor productivity, a Japanese research institute has reported that the Japanese auto industry in 1978 was less productive than the U.S. industry in 1977, but that the degree of difference depended on the sector. Thus, the Japanese assemblers were about 92 percent as productive as U.S. OEMs, but Japanese suppliers had only attained 66 percent of the value-added productivity per worker of U.S. suppliers. In fact, using a value-added definition of labor productivity, the U.S. industry, as of 1980, still exceeded the productivity of the Japanese industry according to both the Japan Productivity Center and the American Productivity Center. Only in 1981 did the U.S. industry fall behind—to 99 percent the value-added productivity of the Japanese industry.

It is somewhat speculative to jump from a value-added measure of productivity to a unit measure. The differences in the value-added measure between the sectors of the two industries may reflect differences in the profit structures of the two industries. Nonetheless, these results, combined with the disparity in U.S. and Japanese reports on unit productivity, suggest that productivity differences may be less at the supplier than at the OEM level. There certainly are implications here for U.S. OEMs who follow a decreased vertical-integration strategy as to whether that strategy necessitates offshore sourcing or whether outside purchasing is sufficient.

If we expand our consideration to include capital productivity, some interesting results emerge. A study of automobile manufacturers by the Japan Productivity Center found that while unit labor productivity at Toyota and Nissan exceeded GM in 1977, GM's total productivity was 16 percent and 58 percent higher, respectively, because its capital productivity was about three times as high. These results, consistent with analyses for 1979 by the Industrial Bank of Japan, suggest that the Japanese edge in unit productivity partially reflects a substitution of capital investments for labor investments. While this seems obvious, a number of industry observers have chosen to ignore or even deny this aspect and focus exclusively upon organizational or management aspects of productivity. A recent U.S. analysis, however,

---

[30]W. J. Abernathy, K. B. Clark, and A. M. Kantrow, *Industrial Renaissance* (New York: Basic Books, 1983), p. 61.

disputes this and argues that capital productivity is also higher in the Japanese industry.[31] One must be cautious, then, in taking these reports or the role of productivity differences in estimates of manufacturing costs as either comprehensively assessing productivity or as applying throughout the entire industry.

In summary, productivity comparisons between two industries are likely to be influenced by the measure of productivity employed, as well as the specific segments of the industry selected for analysis. The variations in reported productivity differentials among the reports discussed in this chapter are indeed quite large, but so, too, are the variations among those reported by ancillary sources. Certainly, extreme caution, as well as careful attention to details of data sources, measures, and manipulation, is indicated in using the information from these reports.

Wages. Compensation differences between the two industries is one of the knottiest areas to tackle. In the first place, management and labor have two quite different perspectives on how the issue should be framed. The worker, quite understandably, is interested primarily in the sum total of benefits received, or overall economic well-being. The manager, equally understandably, is much more concerned with what it costs to provide the sum total of benefits the worker receives. It is not surprising, then, that this issue has been a point of considerable friction within the U.S. industry, especially since the public reports focus upon blue-collar wage levels.

As with the discussion on productivity, Japanese sources present us with a different view of the extent of wage differentials in the two industries. The hourly wage estimates used in these U.S. reports suggest that the wage of the Japanese auto worker is about 53 percent that of his U.S. counterpart. But estimates from a number of Japanese sources suggest that the wage of the Japanese auto worker may well be about 70 percent of the wage of the U.S. worker. Again, it appears that at the assembler level the proportion of Japanese wages may be a bit above this, and in the supplier sector, a bit below.

This discrepancy between Japanese and U.S. estimates of the wage differential is due to a number of problems with our estimates of Japanese wages. The Japanese worker is salaried, receives pay in the form of a variable bonus, and receives many benefits in kind. Most Japanese companies provide some form of housing subsidy—either housing, savings and mortgage plans, or a direct subsidy. Yet these costs are often reported as simply administrative and maintenance costs, or, in a number of cases, not reported. Exactly how other payments in kind are accounted for and costed out is unclear. Some wage costs—for example, provided housing, company buses, recreational facilities, and resorts—may in fact show up in capital accounts. Thus, some

---

[31] Ibid.

reported differences in capital productivity between the industries may be exaggerated and based upon underestimates of Japanese wage costs. Through interview and documents secured at an affiliate member of the Japanese Auto Workers (Jidōsha Sōren) and one of the Japanese manufacturers, the Project was able to develop an estimate of Japanese wages (Working Paper 21). The bulk of the figures are for 1983, although some reflect 1981 costs. It is still the case that there are a number of benefits evaluated according to conservative costing conventions, as well as some benefits for which no cost estimates were available. The estimates themselves, therefore, are conservative. At ¥240 to the dollar, the Japanese auto worker at one manufacturer conservatively makes $11.34 per hour, or roughly 54 percent of the estimated $21.00 for the U.S. worker. (Note that at ¥215, this would be 60 percent, not the 53 percent these reports average.)

Perhaps the most striking result of these comparisons is that the ratio of Japanese to U.S. wages depends very much on the category considered. This is because the Japanese worker receives approximately 82 percent of total compensation in the form of cash wages, while the U.S. worker receives from about 60 percent to 74 percent in this form, depending upon the OEM. Consequently, the Japanese workers receive 73 percent of the average direct compensation received by workers at two U.S. OEMs, but the costs of their indirect compensation are much lower—34 percent of the U.S. costs (see Working Paper 21 for a detailed analysis).

If we examine the benefit received by the worker, we can estimate the comparison of direct wages, reflecting the different methods of providing housing. If the Japanese worker were provided housing through a strict market mechanism, the direct wage would rise to $10.12, or 80 percent of the U.S. worker's. Similarly, if the U.S. industry provided housing at the direct cost level of the Japanese industry, the U.S. workers direct wage could fall to $11.50, raising the ratio to 81 percent. Needless to say, the U.S. worker's cash wage also reflects payments for the cost of living adjustments (COLA), which are the product of inflation. These alone were estimated to cost an additional $1.74 on the 1979 base wage by November of 1981.[32]

The fringe-benefit packages reflect the bulk of the cost difference in labor compensation. Here, too, factors external to the industry play a major role. The Japanese company paid about 34 cents an hour to cover National Health Insurance plus supplementary medical care. The costs in this area for the U.S. industry range anywhere from $2.00 to over $3.00 hourly, depending on the company.

To the extent that these U.S. reports overestimate wage differences, some may underestimate differences in production time, and others may overestimate the total difference. Again, the relative role of the differences in

---

[32]*Automotive News* (9 November 1981), p. 46.

productivity and wage rates is a key problem, and one's answer is strongly affected by the assumptions made about comparative hourly wages in the two industries. On balance, even though these reports probably overestimate the differences in wage levels for blue-collar workers, these differences constitute a substantial source of the MCD. Furthermore, the emphasis upon wage differences for blue-collar workers has been somewhat misleading. While very large differentials in the compensation of very top executives in the two industries have been recognized, these have not been treated as important contributors to the MCD. This is because these amounts are small when averaged over the entire work force. It turns out, however, that differences in the average compensation for white-collar workers in the two industries is itself a substantial contributor to the MCD.

If we rely upon estimates of total labor costs from AHH, it turns out that roughly half of the wage-rate differential is due to blue-collar differences, while the other half is due to white-collar differentials (see Working Paper 21 for details of this analysis).

To a certain extent, wage differences and their associated costs reflect social choices and experiences. The different experiences of inflation between the two societies—reflected in the United States, until recently, in accelerating COLA costs—and the different choices on how to provide medical care have clear implications for the level of direct costs the companies must bear to provide equivalent levels of benefits. So, too, the loss of economies of scale in providing cash rather than in-kind compensation and the need to provide high levels of white-collar compensation to attract and retain these employees in an open market both raise the costs of compensation in the United States. In most, if not all, instances, the U.S. companies are at a serious disadvantage because of these differences.

Alternative Estimates of the MCD

Table 10 (from Working Paper 19) allows the reader to estimate the labor-content cost component of the MCD for a variety of assumptions about comparative wage rates and productivity. Other factors would have to be added or subtracted from this portion to arrive at an overall MCD estimate. The bulk of the information presented in this paper suggests that it is not unlikely that the Japanese manufacturers can produce a small car for roughly half the labor-content costs required by the North American producer. This figure is, of course, quite volatile since it reflects exchange rates, capacity utilization, and productivity improvements in the two industries. At the same time, this estimate of roughly half the labor-content cost in Japan assigns less of a role to cost differences in labor content than the published reports, which often imply that the Japanese labor content is 25 percent or less that of the U.S. industry. If we add other advantages and disadvantages to labor content,

Table 10

Japanese Labor-Content Cost Advantage as a Proportion of U.S.
Labor-Content Cost

(based on differing assumptions about comparative productivity and compensation rates*)

Ratio of Japanese Productivity to U.S. Productivity	Ratio of Japanese to U.S. Compensation Costs										
	.50	.55	.60	.65	.70	.75	.80	.85	.90	.95	1.00
1.00	.500	.450	.400	.350	.300	.250	.200	.150	.100	.050	0.000
1.10	.545	.500	.455	.409	.364	.318	.273	.227	.182	.136	.091
1.20	.583	.542	.500	.458	.417	.375	.333	.292	.250	.208	.167
1.30	.615	.577	.538	.500	.462	.423	.385	.346	.308	.269	.231
1.40	.643	.607	.571	.536	.500	.464	.429	.393	.357	.321	.286
1.50	.667	.633	.600	.567	.533	.500	.467	.433	.400	.367	.333
1.60	.687	.656	.625	.594	.562	.531	.500	.469	.437	.406	.375
1.70	.706	.676	.647	.618	.588	.559	.529	.500	.471	.441	.412
1.80	.722	.694	.667	.639	.611	.583	.556	.528	.500	.472	.444
1.90	.737	.711	.684	.658	.632	.605	.579	.553	.526	.500	.474
2.00	.750	.725	.700	.675	.650	.625	.600	.575	.550	.525	.500

*Cell Entry = $1 - \dfrac{JWR/USWR}{J\ Prod/US\ Prod}$

it is not unreasonable to argue that the MCD at ¥240 is somewhat in excess of $1,500. Again, the major point is that the labor-content costs appear to be less of a factor, and material costs, for example, more of a factor, than these public reports would suggest.

Compensation Systems

The narrow focus upon the cost of compensation has been unfortunate for a number of reasons (Working Paper 20). First, it has become a source of friction within the industry. Management and union have often disagreed on who is to blame. The public wrangling this has occasioned has led much of the public to react to the debate with an attitude of "a plague on both your houses." Second, the narrow focus upon blue-collar compensation costs both reflects and encourages the short-term orientation of corporate decision making in the United States.[33] The issue of compensation costs will be with us for a long time, and it is critical that managers understand and address its long-term and short-term implications. Third, to the extent that compensation costs are viewed as a major and intractable source of the differential in manufacturing costs, the classic response of moving production offshore to cheaper labor markets becomes virtually irresistable. That, of course, will mean a major loss of jobs from the North American industrial base. Fourth, the compensation system for production workers in the Japanese automotive industry is itself a source of long-term competitive advantage. This point has been missed in the debate about the specific difference in the cost of labor compensation.

The Japanese compensation system for the auto worker is quite different than the U.S. system. Payments in kind are much higher, and the worker is salaried and, in some areas, probably experiences a very high ratio of benefits to company cost. The payments in kind and the direct provision of services serve to tie many of his activities directly to the company. The U.S. worker, on the other hand, simply collects hourly pay and goes out into the marketplace. The repetitive tying of benefits to the company, the level of benefits (as in the case of the bonus) reflecting company performance, and the greater encapsulation of the worker's life within the company (e.g., company housing and leisure-time consumption) increase the commitment of the worker and his or her on-the-job motivation to perform. These exist within a system of permanent employment and an organizational and management structure that further promotes the identification of the worker with the company and releases the worker from the fear that increased effort will result in losing jobs.

---

[33]R. L. Banks and S. Wheelwright, "Operations vs. Strategy: Trading Tomorrow for Today," *Harvard Business Review* (May-June 1979), pp. 112-20.

The introduction of profit sharing in the most recent contracts of two U.S. manufacturers, to be sure, is a step in this direction. But it is only one element of a tightly developed system in Japan, and it represents a first and perhaps tentative step in the United States.

The structure of the Japanese compensation system is an important aspect of Japanese organizational life, and even if the compensation costs in the United States and Japan were identical, the perceived benefit by recipients and the actual benefits to the companies would, in all probability, still be higher in Japan. American companies, then, face an area of competitive disadvantage that reflects the preferences of members of our society for arm's-length relationships and a rejection of the old company-town concept, as well as some organizational decisions about compensation form. These will likely remain and undoubtedly continue to be reflected in the specific cost structure of the two industries, and they have competitive implications over and above the actual level of cost differences at any particular moment.

Implications

First, in light of the variations of the differences in manufacturing costs among these public estimates, it is less than clear what the total cost difference is, how it may vary over sectors of the industry, or how it might fluctuate with changing circumstances. That it is substantial and of extreme competitive importance, however, is clear.

Second, and more importantly, it is less than clear what the role of many factors in the cost difference may be. Our understanding of these factors must be sharpened in order to develop appropriate responses. In particular, what is needed is a clearer picture of the relative contribution of productivity and wage differences, compared to other factors as well as to each other, to the overall cost difference. Partial measures of productivity and estimates of Japanese wage costs may obscure and distort overall comparisons and result in inappropriate assessments of their relative contribution.

The assessments of wage and productivity factors made in these reports have been a source of friction not only between union and management but also between manufacturing and financial managers. This friction can prevent concerted efforts to address the problem, and the lack of clarity may obscure proper remedies. For example, if wages are the major source of the cost difference, is the effort to learn more from the Japanese about management and manufacturing techniques misplaced? If productivity is the major source, then the current emphasis on rolling back wage rates, even temporarily, may be counterproductive since in real wages the Japanese appear to be closing the gap. It may only serve to heighten acrimony in the United States between the union and companies. If productivity is the real source, what are the roles of manning levels, inventory systems, scrap and rework, and patterns of job assignment? To be sure, the differences are significantly large so that the American producers need to be working on all these factors. To the extent that portions of the MCD arise from social choice and government actions, the manufacturers are faced with problems outside their control. The very best efforts of American producers will not completely eliminate the current MCD.

Third, while the proprietary studies conducted by the U.S. auto producers may be completely accurate, we cannot assume their general applicability throughout the industry without knowing what factors were considered or what assumptions were made in developing them. One company's problems are not necessarily another's. The variability they themselves reveal should alert us to proceed with caution.

Fourth, to the extent that some differences in manufacturing costs reflect societal choices and preferences, these issues must be examined from a different perspective. Blaming the industry for bad management and high labor rates will not resolve these issues.

Chapter 8
Product and Process Evolution

Product Evolution

Introduction

The international automotive industry is experiencing a period of dramatic change. Essentially every facet of the industry is undergoing a metamorphosis. This is true of both product and manufacturing technology. Viewed casually, the product itself may appear to be undergoing little real change or shift, but this is only true of outward appearance. While the vehicle is evolving at a modest pace as a total system on the macroscale, a whole series of microscale revolutions are occurring within the subsystems of the vehicle. Dramatic changes in the use of electronics, materials, and new processing techniques are but a few of the areas of revolution.

It is clear that the United States, Japan, and the remainder of the auto-producing areas of the world are making rapid technical progress. Detailed innovation is perhaps moving at a faster pace than ever before in the history of the auto industry. Certainly, this is in part prompted by the increasingly competitive nature of the industry. It is evident further that the automotive industry will retain its high level of importance, at least for the foreseeable future, in the United States, Japan, and Europe. Consequently, the technology of the automotive industry in terms of both manufacturing and product technology represent important national assets. In fact, the spirited competitive environment and the rapid technological changes in the automotive industry are spawning dramatic change in other industries as well. Robotics, for example, as applied in heavy industry, is receiving its primary stimulation from automotive manufacturers and suppliers. Even the engineering systems of the auto industry, such as computer-aided design (CAD), are being applied to other industries. Pioneered in the aerospace industry, CAD has been advanced by the automotive industry to the point where it holds state-of-the-art leadership and is building a technical base and capability that has many applications throughout the world.

American and Japanese studies indicate there are important quality differences between U.S. and Japanese products. For example, the United States is viewed as superior in areas such as corrosion and safety, while Japan appears to be better in such areas as fit and finish quality and total car reliability. A number of areas are not viewed as being significantly different.

In general, however, all of these quality areas that reflect value judgments on the part of the consumer may approach equilibrium in the future. That is to say, quality differences between U.S. and Japanese automobiles and trucks are diminishing and may continue to diminish, though, as we shall see in the results of our survey of Japanese suppliers in chapter 9, the continued stress on quality improvement in Japanese second-tier suppliers makes them a "moving target." Notwithstanding, other factors besides quality will be stressed in seeking competitive advantage, and one of these, in the opinion of numerous experts, appears to be advances in automotive technology.

The role of the automotive supplier in the technological revolution of the automotive industry is particularly important. It is evident that vehicle manufacturers will be relying more on the suppliers than in the past and attempting to forge stronger relationships. This is particularly pertinent in the United States, where manufacturer-supplier relationships have been generally at arm's-length. This trend should stimulate supplier creativity and innovation on an international scale. It is important to recognize the magnitude of the stimulus provided by the automotive industry because of its physical size and dollar volume. At the same time, the continued pressure from manufacturers for cost reduction means that the reward for a successful supplier that displays the ability and innovation to make the transition from customer-designed to codesigned products is to "keep the business." That is to say, such transitions are becoming a condition of doing business with the manufacturer rather than a guarantee of enormous profits.

Products being designed today for sale tomorrow can only be successful if the market will accept them. Consequently, the automotive industry must work diligently to design and produce vehicles that match often volatile customer desires. This is a complex problem that probably will continue because of rapidly shifting external factors that influence consumer preferences, which, in turn, lead to shifts in the relative importance of purchase criteria. For example, the two oil shocks of the 1970s led, temporarily, to rapid shifts to small vehicles in the United States. In Japan and most of the remainder of the world, high fuel prices had already dictated concentration on smaller cars. In effect, we have seen a narrowing of the gap between the characteristics of cars sold in the United States and foreign countries. However, the fuel price differential between the United States and most of the developed world remains significant. Delphi forecasts for 1992 diesel and gasoline retail fuel costs in the United States are $1.90 and $1.75 per gallon, respectively. In Japan, the retail gasoline price forecast for 1992 is equivalent to $3.65 per gallon, and $3.01 for diesel fuel. Today, the general flattening in the trend of fuel prices in the United States has led consumers to reassert interest in larger vehicles. In spite of this recent trend, however, U.S. manufacturers will remain on a course of downsizing and weight reduction, although the average domestic car is likely to be somewhat larger and more powerful than those produced and sold in Japan.

Important differences among nations in consumer requirements and such factors as emission regulations and safety standards will lead to important technological differences in the products being used in various countries. For example, in Europe, without rigorous emission standards and great emphasis on performance, the specific output of engines produced there are much higher than in the United States and Japan. Of course, the Japanese in particular are keenly sensitive to the U.S. market and export vehicles tailored to U.S. requirements.

Some differences in the characteristics of U.S. and Japanese vehicles are expected. For example, the price (in current dollars) of the average vehicle sold in the United States in 1992 is expected to be $11,250, whereas in Japan the average price in equivalent U.S. dollars is expected to be $6,250. In 1992, the average life of a passenger car is expected to be thirteen years in the United States and ten years in Japan. Marketing experts in Japan suggest that the average car size in Japan will continue to be significantly smaller than in the United States.

It is clear that the emphasis on both product and manufacturing technology in the United States and Japan, as well as the other producing areas of the world, will maintain investment requirements at a high level. This is certainly characteristic of any industry in transition. The greater financial risk is in not taking risk or not making investment in new products and processes. If they do not invest, manufacturers and suppliers will risk losing a competitive position. It is also important that government policy affecting the industry be structured to ensure that the climate and pace of investment is adequate to address the requirements of the technological revolution.

Trends in Automotive Product Technology

Delphi studies conducted by both the U.S. and Japanese research teams in their respective countries have provided valuable insight into automotive technology predictions through the year 1992.[34] The Delphi survey technique is a systematic, iterative forecasting method based on independent data from individual experts in both the United States and Japan. The technique encourages, where warranted, a consensus among the panel of experts regarding future events but does not suppress equally important differences of opinion. The Delphi method relies on expert judgment and recognizes that forecasts on which decisions regarding policy must rely are influenced by personal preferences and expectations, in addition to more quantitative factors.

---

[34]Except where noted, the data in the following sections of this chapter are from Delphi surveys made in 1982-83.

Panelists in the Japanese Delphi study consisted of 50 experts from Japanese vehicle manufacturers and 20 from prominent suppliers. In the U.S. survey, more than 200 engineering and marketing managers and executives responded. Of these, about 30 percent were employed by vehicle manufacturers, 40 percent by suppliers of component parts, and the balance by suppliers of materials and equipment. Panelists were identified experts on various aspects of automotive technology, and some are in positions of sufficient authority as to enable them to make their forecasts come true. The surveys focused on vehicles *produced* in each country and not on all vehicles sold in the United States and Japan. Furthermore, since a large fraction of cars produced in Japan are exported, the Japanese data are not purely reflective of the domestic Japanese market.

In this report, selected technical trends include: total vehicle characteristics, powerplants, drivetrains, electronics, accessories, and materials. In addition, several forecasts of business factors are presented. The U.S. and Japanese forecasts are presented separately, followed by comparisons of the results.

U.S. Automotive Trends

Total Vehicle. The average weight of U.S.-produced passenger cars is expected to decline from 3,300 pounds (1,495 kg) in 1981 to 2,500 pounds (1,135 kg) in 1987, and 2,250 pounds (1,020 kg) by 1992. Continued gains in fuel economy are forecast to result from technical advances. Clearly, even with reduced expectations for U.S. energy prices, fuel economy remains one of the dominant factors influencing automotive designs. Some moderation in the fuel economy target is noted in comparison to two years ago, and much less emphasis is placed on downsizing as a means to achieve efficiency gains. Factors expected to contribute to improvement in fuel economy, listed in order of contribution by 1992, are: engine efficiency (24 percent); drivetrain and transmission (24 percent); lighter-weight materials (18 percent); downsizing (12 percent); aerodynamics (12 percent); tire-rolling resistance (7 percent); and reduced performance (3 percent).

Materials. The use of high-strength, low-alloy (HSLA) steel will increase as the mass of low-carbon steel and cast iron are reduced. However, the prospects for both aluminum and plastic components are somewhat reduced from the previous Delphi, although they both remain as important automotive materials. Panelists forecast significant gains in corrosion performance.

Engines and Drivetrains. The shift to smaller engines with fewer cylinders is expected to continue. The predominant engines in 1992 will be the in-line four-cylinder (58 percent in 1992) and the V6 (32 percent in 1992). Some three-cylinder engines are forecast; V8 engines are expected to decline to 5 percent of U.S. production. Diesel engines are expected in 10 percent of U.S.-produced cars by 1992. Optimism for the diesel was far less than noted in the prior Delphi study, in which the diesel was forecast in 20 percent of passenger cars in 1990. Battery-powered electric vehicles are not expected to become commercial before the 1990s.

The Dephi forecasts for the range of displacement of gasoline and diesel passenger-car engines are shown in table 11. The efficiency of both diesel and gasoline engines is forecast to improve by 15 percent over 1980 designs, according to an earlier (1981) Delphi study.

The use of the front-engine/front-drive configuration, with the engine located transversely, will continue to expand rapidly, although not to the extent predicted two years ago. Forecasts call for an 81 percent utilization of this configuration in passenger cars by 1992, with the remainder front engine/rear drive (15 percent) and mid-engine/rear drive (4 percent).

Electronics. The use of electronics will continue to expand. Microprocessors, for example, are expected in essentially all of the future U.S.-produced passenger cars and light trucks. By 1992, the fraction of total vehicle cost represented by electronic componentry, such as microprocessors, transducers, and actuators, is expected to rise to 12 percent. In more detail, electronic components in 1992 are forecast to represent 10 percent of the cost of safety systems, 15 percent of the drivetrains, and 40 percent of comfort and convenience systems.

Accessories. The use of power brakes and power steering are expected to be essentially the same as at present. On the other hand, air conditioning is expected to expand considerably in the next ten years (based, in part, on the 1981 U.S. Delphi report).

Business and Management Factors. Panelists forecast (in the 1981 Delphi survey) that factors likely to hinder technological progress are, in order of descending importance: requirements for capital investment, economic conditions, availablity of machine tools, conversion time, and availability of engineering and technical labor.

A 20 percent increase in the use of "off-line" subassemblies is forecast for manufacturers by 1992. Among the component types subject to off-line subassembly, electronic components were most prominently cited, followed by wheels, brakes, instrumentation, and seat assemblies. Numerous other subsystems were noted as well.

One of the most provocative Delphi findings was the forecast of a major increase in offshore sourcing of components by the manufacturers. These are presented in table 12, in terms of dollar volume.

Japanese Automotive Trends

Total Vehicle. In the category of compact cars, such as the Corolla and Sentra, the average vehicle weight will decrease 1-2 percent annually.

Table 11
Forecast for the Range of Displacement
of Gasoline and Diesel Passenger-Car Engines
(in percentage)

Spark-ignited Engine Displacement-Liters	Median Forecast		Diesel Engine Displacement-Liters	Median Forecast	
	1987	1992		1987	1992
Over 5.0	5.0	2.0	Over 5.0	11.0	6.0
3.0 - 5.0	37.0	30.0	3.0 - 5.0	56.0	50.0
1.5 - 3.0	53.0	58.0	1.5 - 3.0	33.0	44.0
Below 1.5	5.0	10.0	Below 1.5	0.0	0.0
TOTAL	100.0	100.0	TOTAL	100.0	100.0

Table 12
Forecast of Offshore Sourcing of Components
(in percentage of dollar volume of parts sourced outside U.S.)

Outside Parts Source	Median Forecast 1987	1992
Canada	8.0	8.0
Western Europe	3.0	5.0
Japan	6.0	10.0
Other Asia	2.0	4.0
Mexico and South America	5.0	7.0
Overall Total	24.0	34.0

Improvements in engine efficiency are forecast to contribute substantially to vehicle fuel economy, followed by powertrain improvement, weight reduction, and reduced driving resistance.

Materials. For weight reduction, high-strength low-allow (HSLA) steel will be the leading material. The use of plastics will also increase over the next ten years. It is difficult to increase the use of aluminum because of its cost penalty.

Engines and Drivetrains. Three out of four passenger cars will continue to be equipped with four-cylinder engines through 1992. Six-cylinder engines, mainly used in luxury models, will account for 15 percent. In the minicar category, there will be increased use of three-cylinder engines. Diesel engines are forecast in 10 percent of passenger cars and 50 percent of light trucks by 1992.

Displacement will be less than 2.0 l in 98 percent of passenger cars. Average displacement will decrease slightly as the 1.5 l and smaller engines increase their share at the expense of the 1.5-2.0 l group. Engines less than 0.55 l will account for a 3-4 percent share of passenger cars.

Engineering development is expected to improve efficiency by 10 percent in five years and 15 percent in fifteen years in both gasoline and direct-injection diesel engines, mainly through the use of ceramic materials with high heat resistance.

The front-engine/front-drive system is forecast to increase its share of the passenger-car market, from the present 20 percent level to 50 percent in five years and 65 percent in ten years.

Electronics. Microprocessors are forecast in 30 percent of passenger cars in five years and 60 percent in ten years and, in the same time frame, in 10 percent and 20 percent of light trucks. The distribution of electronic applications will be 50 percent (cost basis) in engine and powertrain, 20 percent in entertainment, 15 percent in safety devices, and 15 percent in air conditioning and related areas. The forecast of engine and drivetrain electronics allocates 30 percent to the microprocessor and 45 percent to sensors and actuators.

Accessories. Demand for driving ease is forecast to increase the installation of automatic transmissions to 50 percent in ten years. The continuously variable transmission is forecast at a 5 percent market share in ten years. Power brakes, power steering, and air conditioners are all forecast to show higher installation rates.

Business and Management Factors. In passenger-car production, 81 percent of the panelists expect that off-line subassemblies by suppliers will increase over the next ten years. Subassemblies will include electric parts, seats, dashboard/instrument panels, and air-conditioning units. Offshore procurement of parts is forecast to increase to 8 percent in five years and 14 percent in ten years, and the United States is expected to be the leading source among foreign countries.

Comparison of Japanese and U.S. Forecasts

Total Vehicle. In the United States, a rapid decrease in vehicle weight is forecast; in Japan, the rate of decrease is expected to be substantially less— only 1-2 percent per year—because the Japanese cars are already much smaller, on the average, than those in the United States. Technical factors expected to improve fuel economy are about the same in the United States and Japan, with both placing greatest emphasis on improved engine efficiency. Lower-weight materials will be reasonably popular in both countries. Both the United States and Japan are expected to place greater reliance on high-strength low-alloy steel in their weight-reduction programs, but the United States will continue to use more aluminum and plastic. A detailed comparative presentation is shown in table 13.

Delphi experts in both countries generally concurred on the relative advantages and disadvantages of basic materials, although there were some exceptions. For example, mild steel formability received greater support in Japan as a positive factor, and corrosion was less important as a negative consideration, although corrosion control is important in future designs in both countries.

Engines and Drivetrains. While U.S. manufacturers will downsize engines and reduce the number of cylinders, the Japanese forecast little change in their present engine mix since their current average engine size approximates foreseeable demand. The diesel engine is expected to play a more prominent

Table 13
Forecast of Material Usage
(in pounds per car)

	U.S. Median Forecast		Japanese Median Forecast	
	1987	1992	1987	1992
Low Carbon Steel	1300 *	1060 *	913 *	673 *
HSLA Steel	300 *	320 *	297 *	427 *
Total Steel	1600	1380	1210	1100
Total Cast Iron	360	250	165	143
Total Zinc	15	15	11	11
Aluminum Castings	102	102	47 *	55
Wrought Aluminum	35	33	19 *	22
Total Aluminum	137	135	66	77
Plastics				
Nonreinforced	110 *	100	77	88
Fiber Reinforced	125 *	135	22	33
Metal Laminate	15 *	25	11	11
Total Plastic	250	260	110	143
Total Copper	22	22	15	11
Total Magnesium	2	3	2	4
Total Glass	80	75	66	55
Total Rubber (Excl. Tires)	35	30	33	33

*Raw data adjusted to match subtotal. Preliminary results for U.S. data.

role in both countries, although not as large in the United States as once expected. Japanese cars already achieve high fuel economy but will use approximately the same fraction of diesels (10 percent) as the United States in 1992. In 1992, there is a far greater expectation for diesel use in light trucks in Japan (50 percent) than in the United States (25 percent), particularly in commercially owned vehicles where fuel economy is a major concern. It is clear that regulations for exhaust emission will exert a powerful influence on diesel use in both countries. There are even lower expectations for electric vehicles in Japan than in the United States. Japanese and U.S. experts expect similar improvements in engine economy due to advances in engines and drivetrains in the decade ahead.

Reconfiguration of passenger cars to a front-engine/front-drive configuration will be less complete in Japan than in the United States. In 1992, the United States is expected to produce 81 percent front-drive passenger cars, whereas Japan is expected to produce only 65 percent. In Japan, front-wheel drive has been associated with very small, inexpensive cars, so an image and marketing problem exists. Transmission expectations differ between the United States and Japan, as shown in table 14.

Japan will continue to use a higher percentage of standard transmissions than the United States, but significant growth in the use of automatic transmissions is forecast. An important technological development in both countries will be the appearance of continuously variable transmissions later in the 1980s.

Electronics. Japanese-produced cars and trucks will make significantly less use of microprocessors than in the United States. This is perhaps a reflection of major differences in the basic strategy of manufacturers, owing in part to lighter-weight and smaller Japanese cars that require less technical sophistication to meet demands for emission control and fuel economy. Also, the cost of this technology in relation to the basic vehicle may be a consideration. On a basis of cost allocation, both the United States and Japan believe electronics will be distributed similarly through the vehicle. Both countries expect a significant increase in electronic sophistication and new innovations, including detailed traffic information, diagnostics, etc. The cost allocation for engine electronics is similar between the United States and Japan.

Accessories. Significant differences are forecast. In Japan, power-assisted and comfort items will increase significantly. In the United States, the rate of growth will be smaller because of an already high level of application of these items.

Business and Management Factors. Capital investment is viewed almost equally in the United States and Japan as the most important factor limiting conversion to new technology. Both countries will experience large increases in off-line subassembly use, but the Japanese are expected to be much less aggressive in the pursuit of offshore sources for automotive components.

Table 14
Forecast of Transmission Usage in the U.S. and Japan
(percent of total transmissions)

	United States		Japan	
	Median Forecast			
	1987	1992	1987	1992
Manual				
Three Speed	0.0	0.0	1.0	1.0
Four Speed	15.0	10.0	29.0	16.0
Five Speed	10.0	16.0	30.0	33.0
Total Manual	25.0	26.0	60.0	50.0
Automatic				
Three Speed	36.0	19.0	20.0	15.0
Four Speed	36.0	41.0	18.0	30.0
Continuous Variable	3.0	14.0	1.0	5.0
Total Automatic	75.0	74.0	40.0	50.0
Total	100.0	100.0	100.0	100.0

Summary

The U.S. and Japanese automotive industries will continue to face dynamic changes. However, the U.S. industry will experience a higher rate of change, particularly in major capital-intensive areas such as engine, drive-train, and basic reconfiguration in downsizing requirements.

Process Evolution

The rapid product evolution that has occurred in the U.S. automotive industry, and which is projected to continue in both the Japanese and U.S. industries, has a direct impact on the evolution of auto-manufacturing processes. The forces that drive this product evolution influence the choice of manufacturing technologies in different ways. In the United States, these forces include the federal requirements for improved fleet fuel economy, federal emission and pollution regulations, federally mandated safety standards, competition from imported vehicles based on quality, lower cost, and design features, and the depressed U.S. market for automobiles.

The simultaneous occurrence of these forces has necessitated a fundamental restructuring of the manufacturing processes used by the U.S. industry, with capital expenditures over the past five years rising to a record $51 billion. The process changes initiated by the U.S. firms differ in degree and complexity, depending upon the firms' individual strategies, resources, and prospects. Furthermore, the manufacture of automobiles is a multi-stage process, roughly diagrammed in figure 8. Each stage is comprised of highly diverse and distinct manufacturing processes. For example, the fabrication of raw materials into parts and components draws on the technologies of steel making, tire manufacturing, glass making, malleable and grey iron casting, plastic molding, and soft-goods fabric weaving, to cite a few. The automotive industry is so large that it consumes major fractions of the capacity of these process industries. Therefore, any change in the automotive product that alters or eliminates the automotive demand for such materials may have major implications for employment and economic policy both in the United States and in Japan. Most of these process industries are considered part of the supplier industry. Also, within this supplier industry, manufacturers of fabricated components (fasteners, stampings, a large variety of component subassemblies such as instruments, radios, windshield wipers, shock absorbers, spark plugs, pumps, etc.) comprise an additional diverse and significant element of the automotive industry. Within the manufacturing process itself, however, the fabrication and assembly of engines and transmissions, the stamping and assembly of sheet-metal body parts, and the assembly of the final vehicle are the most distinct and important manufacturing stages. In discussing changes in process technology within automotive firms, therefore,

FIGURE 8

Overview of Typical Passenger-Car Manufacturing Stages

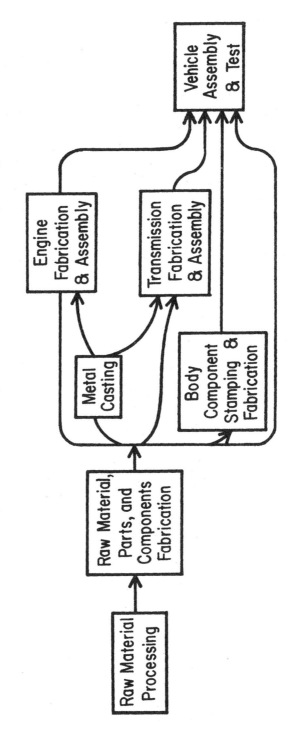

we must recognize that the potential for technological and strategic evolution differs among these four stages. Furthermore, because of the historical differences between the process strategies in the United States and Japan, the two industries face some unique process alternatives, as well as some common ones.

Final-assembly plants for major automotive manufacturers utilize basically similar technologies throughout the world. Typically, there are four major stages within an assembly plant: body assembly, finishing and painting, body trim, and final assembly. Both U.S. and Japanese assembly plants assemble a mix of body styles, with options such as engine size, manual versus automatic transmission, front- or rear-wheel drive, and left- or right-hand steering on the same final-assembly line. In addition, each vehicle may have any one of a large combination of optional accessories such as air conditioning, tape deck, wheel covers, alternate upholstery, seat types, etc. Thus, the typical final-assembly line builds a mix of vehicles in sequence using a fixed cycle. Because of the variety and complexity of the product itself, the final-assembly and trim lines are labor intensive. While robots and similar automation have been applied to a few isolated operations, such as positioning batteries or seats for the assembler to install, an assembly line producing 60 vehicles per hour may use up to 600 assemblers per shift. Little major innovation has occurred in the efficiency of U.S. trim and final-assembly lines in recent years.

By contrast, both the paint and the body-assembly stages have changed dramatically in the last decade. Paint finishes are far more corrosive resistant, as well as better looking. Most, if not all, of the human operators have been replaced by dip tanks and laser- or robot-controlled spray operations. Rust resistance is greatly improved, and quality is higher and more uniform. In body assembly, manually operated spot-welding guns have been replaced with robotic spot welding and automated clamping fixtures for achieving a reliable fit of body parts prior to welding. Up to 95 percent of spot welds are performed by robots in the modern assembly plants. Technically, this operation is now amenable to full automation.

Automotive manufacturing capacity is usually reported in terms of the number of jobs per hour that can be produced on the final-assembly line. Estimates of the assembly capacity of U.S. domestic manufacturing plants are given in table 15. Clearly, the U.S. industry has excess assembly capacity in the aggregate. However, since assembly plants are tooled for specific car lines, a shortage of capacity can arise if demand is strong for a particular car line. Thus, U.S. carmakers face a difficult problem in configuring assembly capacity when U.S. demand shifts rapidly from standard-size to compact cars and then back again in less time than the carmakers can retool assembly plants. By contrast, the Japanese manufacturers currently have fewer difficulties in this regard since their car lines are concentrated in compact and subcompact sizes.

Table 15
Capacity of U.S. Final-Assembly Plants
(passenger cars)

	Model Year					
	1977	1978	1979	1980	1981	1982
Number of Assembly Lines	48	49	47	47	43	41
Planned Number of Shifts	84	91	90	73	66	68
Planned Capacity[a] (000s)	8,688	9,549	9,600	9,050	7,321	7,683
Actual Output (000s)	9,104	8,940	9,208	6,776	6,673	NA
Two-Shift Capacity[b] @ 60 Per Hour (000s)	11,520	11,760	11,280	11,280	10,320	9,840

Notes: [a]Planned Capacity = total of jobs per hour times 2,000 hours per shift times planned number of shifts.
[b]Two-Shift Capacity = number of assembly lines times 60 per hour times two shifts times 2,000 hours per shift.

Source: *Ward's Automotive Yearbook*, various years.

Comparisons between final-assembly plants in the United States and those in Japan should be viewed with some skepticism because of the difficulty of generalizing about plants constructed in different years and because of the different size and mix of vehicles. Nevertheless, some qualitative differences in process and management technologies no doubt contribute to the conclusion that Japanese automotive manufacturers achieve lower cost than their U.S. competitors. Among the more generally recognized differences is the smaller vehicle size, on the average, in Japan. Smaller vehicles physically require less plant and tooling space than larger ones. In addition, the close attention in Japan to defect prevention in assembly reduces the floor area required for repair and retest of deficient vehicles. The use of just-in-time delivery of parts and components to the assembly work station allows Japanese assembly plants to carry smaller parts inventories, reduces handling and damage costs, and reduces the plant area used for parts storage. U.S. manufacturers now are fully aware of these differences and are initiating the studies or programs needed to match or exceed the effectiveness of process methods used by the Japanese. However, improvements will not be achieved quickly, nor will they be achieved by imitation of Japanese processes and systems. For example, the long-standing involvement of the U.S. manufacturers with the UAW has created a third party positioned between the worker and management with distinctly different operational consequences than that of the Japanese unions. The U.S. practice tends to limit a worker's flexibility to perform a variety of tasks, to develop permanent work standards, which tend to fix productivity levels, and to permit bumping according to seniority without high regard to skills. It creates a real threat of company-wide strikes and shutdown, produces complex grievance procedures, necessitates the need to work out national and plant local agreements over the bargaining table, tends to separate inspection and routine maintenance from the tasks of the production worker, and uses floating relief rather than mass relief. Because of the large proportion of workers currently involved in trim and final-assembly jobs, improvements in the assembly process will be heavily dependent on the ability of U.S. industry to deal with these human aspects of the assembly technology.

As indicated in figure 8, engine and transmission fabrication and assembly are two major component-manufacturing processes, both from the standpoint of significance to the value of the product and also in terms of the capital cost of the required production plant. Currently, a new engine or transmission plant, designed for a capacity of 400,000 units per year, costs up to $500 million. The typical plant will consist of two main processes: (1) high-precision machining of the various components and parts and (2) assembly and test of the subassemblies and final product. Traditionally, fixed automation, such as transfer lines, has been used for cost-effective machining and grinding of the engine blocks, shafts, pistons, gears, transmission housing, etc. Such automation permits the simultaneous machining of parts at sixty or more work stations, with parts indexing between stations on a cycle of less than one minute. The process minimizes floor space yet provides high precision since the part is palletized or under precise control throughout its passage through the series of operations. Because of their high capital cost and limited flexibility, transfer lines require continuous production over five to ten years to justify their purchase. By comparison, the assembly of engines and transmissions has tended to be labor intensive, using short-cycle, fixed-movement, conveyorized lines. Transmission and, particularly, engine testing have depended upon fixed test stands, requiring operator intervention if the assembly fails any phase of its test cycle.

The trend toward smaller vehicles, with four- or six-cylinder engines instead of V8 engines, front-wheel drive with new transaxle designs, and reductions in weight and emissions, has necessitated retooling or the construction of new U.S. engine and transmission plants. While many of these plants continue the use of transfer-line technology, where product life and anticipated volumes justify the investment, newly developing technologies are modifying the basic strategies for some of these facilities. The principal driving force is the desire to permit economical changes in product design without completely retooling the plants. This need manifests itself in industry efforts to develop flexible manufacturing systems suitable for responding to rapidly changing designs and customer demands. Furthermore, flexible systems are intended to accommodate to the greater variety of designs to be manufactured, each with small volumes and batch sizes that are not economical if produced on dedicated transfer lines for continuous machining. Computer-controlled machining centers (with automated handling systems for routing and scheduling the movement of parts among the machine tools) have been utilized for small batches of parts required for truck and off-road vehicles and service components for obsolete passenger-car models.

The focus on improved quality and flexible manufacturing systems tends to tighten the machining and finish tolerances that are required. Information systems, computer communication systems, and programmable controllers are contributing significantly to improvements in quality and cost reduction in both the fabrication and assembly operations of transmission and engine plants. Adaptive control, which adjusts cutting operations to process changes, matches piston to cylinder dimensions, or adjusts the engine test sequence according to the performance of the engine on the test stand, is an example of improved technology arising from the use of programmable controllers. The technology of controllers and microprocessors also greatly alters the nature of the work force in these plants. Because the assembled engines and transmissions are more compact than finished vehicles, their assembly is more amenable to accomplishment by robotics, with minor use of human operators. Hence, the loading and unloading of machining and test operations and the assembly of final products in engine and transmission plants will be increasingly converted to robotics, with an associated decrease in the manhours required by such facilities. We note that the use of programmable logic will tend to permit the flexibility desired in today's plants even though the fraction of manual operations will be reduced. In this connection, the recent Delphi study reports that its respondents perceive that the Japanese manufacturers are more advanced in the introduction of robots but U.S. manufacturers are more advanced in the use of CAD/CAM technology. By 1992, the forecast strongly projects the two industries will be closely comparable in both technologies. To be sure, robots do not guarantee higher productivity, nor does CAD/CAM guarantee the integration of design and manufacturing. Therefore, how these technologies are used is just as important as their technological sophistication and level of diffusion.

The traditional standard and compact vehicle uses a body fabricated from sheet steel. Typically, the steel is received in coils, blanked, stamped in a sequence of large presses, assembled in fixtures for tack welding, then spot welded into its final structural form. In Japan, most of the stamping and body-welding operations are located in or adjacent to the final-assembly plants. In the United States, in order to achieve economies of scale in both stamping and final-assembly operations and to minimize the shipping costs of finished vehicles to nationally dispersed dealers, stamping plants tend to be geographically separated from and fewer than assembly plants. In time, it is probable that this historical development will change toward more integrated complexes of assembly plants and feeder stamping plants similar to the Japanese practice. Furthermore, the Japanese have made outstanding advances in their ability to make quick die changes in their press operations. The U.S. drive toward flexibility in order to respond to shifts in customer demand, decrease inventories, and improve quality will reinforce U.S. efforts in this technology. Automation in stamping plants will tend to reduce the number of operators as more automated handling systems are installed and robots become the norm for positioning and welding the body components.

Research and Development Processes

The escalating international competition in passenger vehicles has motivated renewed emphasis on R&D within the automotive industry. In 1979, GM, Chrysler, and Ford spent about $4 billion on R&D, compared to a total slightly over $1 billion for both R&D and capital investment by Toyota and Nissan in the same period. In addition, U.S. motor-vehicle and parts manufacturers spent $8.3 billion for capital investment, and, as noted in chapter 2, these high spending levels have been sustained. Expenditures have been designed to radically upgrade outdated product offerings and to replace obsolete facilities. Were it possible for U.S. manufacturers to continue such investment patterns, they might be expected to establish a technological edge over their competition in both product and process technology. Furthermore, the U.S. commitment to and recognized leadership in the development of CAD/CAM systems suggests that new U.S. product designs should become available with shorter lead times and with greater design reliability. Another of the potential benefits of CAD/CAM, as noted above, is the integration of product design with manufacturing-process design. To be sure, the Japanese producers are moving rapidly in this direction as well. A new vitality and experimental approach may characterize the evolution of U.S. auto-manufacturing processes for vehicles of the 1990s. This, in turn, can be expected to trigger renewed competitiveness on the part of Japanese auto producers.

Chapter 9
Manufacturer-Supplier Relations

Introduction

Manufacturer-supplier relations is a subject that has not received the scrutiny and analysis commensurate with its importance to the competitive success of automobile companies. This is in part because of the complexity of the subject and the very diffuseness of the industry. The difficulty of conducting a comparative study of the Japanese and U.S. supplier industries is compounded because of the differing organizational structures. This chapter itself reflects this difficulty. The degree of integration is less than one would desire, but we do believe that in highlighting certain critical areas in each of the supplier industries, not always the same ones, we will have significantly advanced the state of knowledge.

Just how is the supplier relationship critical to the competitive efforts of the automobile manufacturers? We will discuss the matter from a variety of perspectives. One U.S. automobile manufacturer estimates that between 20-33 percent of the cost advantage of the Japanese is supplier related. They attributed this to: smaller and lower-cost facilities, more efficient equipment utilization, less expensive manufacturing processes, and better utilization of human resources. One consequence of this kind of thinking has been a drastic reevaluation of manufacturer-supplier relations in the United States, with close examination of the Japanese model.

We begin the chapter with a characterization of the two supplier industries, which we hope will aid the reader in interpreting the analysis to follow. We then present a summary comparison of the OEM purchasing function in the two countries. Since purchasing is one of the critical links to supplier organizations, this comparison will shed some additional light on how the supplier relationships are organized. A particular focus of this investigation is pricing practices in the two industries and their significance for manufacturer-supplier relations.

We then turn to the results of our survey of first- and second-tier suppliers in Japan. This is an area about which relatively little has been published and considerable confusion, particularly among Western observers, exists. In this section, we will examine the relationships of first- and second-tier suppliers with manufacturers and compare the character of the labor force, wages, and other benefits of first- and second-tier suppliers. We will also analyze the different business strategies of these suppliers.

151

The final section of the chapter will be based on our case studies of window regulators and disc-brake calipers in the U.S. automobile industry. These case studies were designed to provide detailed treatment of the formal and informal practices guiding the relationships with OEMs. Particular emphasis is placed on the implications for technological innovation. Our concern is with assessing whether these OEM-supplier relationships provide suppliers with the incentives and support to play a significant role in technological innovation.

Characterization of the Two Industries

Given the diffuse structure of the supplier industries in Japan and the United States and their varied dimensions, it is exceedingly difficult to characterize the respective industries. For example, data on the value of shipments to OEMs versus servicing the aftermarket are often not disaggregated. The problem of characterization is particularly severe in the United States where the industry is not as organized as its Japanese counterpart.

Based on our research, we estimate that total sales of raw materials, components, and parts by the U.S. supplier industry to U.S.-based manufacturers totaled $40 billion in 1980 (Working Paper 10). This does not include transfers from divisions or subsidiaries to the OEMs. Although there are nominally over 40,000 supplier firms, sales of the 4,800 suppliers doing $700,000 or more in business a year totaled $34 billion to three major automotive manufacturers. Within this group, the sales of only 120 suppliers (primarily suppliers of raw materials) accounted for $18 billion. These amounts, however, are based on 1980 data. With the deepening recession, sales fell still further through 1982. The real value of shipments of domestic parts and component producers declined in 1982 to a level 38 percent below the 1978 peak. In 1979, there were an estimated 1.4 million workers in the supplier industry (including second- and third-tier suppliers who did not supply the OEMs directly). At least one-third of those jobs are estimated to have been lost in the decline that extended until 1983.

It is difficult to compile comparable data for the Japanese industry because of different reporting conventions. The 400 members of the Japan Auto Parts Industries Association (JAPIA) reported for the period 1980-81 a production value of some $19.2 billion (excluding exports). Although there are an estimated 8,000 supplier firms, JAPIA accounts for about 90 percent of the parts and components industry's total output.[35] Based on the 90 percent

---

[35]Dodwell, *The Structure of the Japanese Motor Components Industry* (Tokyo: Dodwell Marketing, 1979).

figure, we arrive at a total production value of $21.3 billion. However, this total does not include suppliers of raw materials. Based on the Japanese Census of Manufacturers, we estimate employment to be about 460,000 (auto parts and components, including auto bodies).

While all these data are quite incomplete and suggestive at best, they do point to the more compact nature of the Japanese supplier industry. Whereas 4,800 U.S. suppliers account for 85 percent of sales, in Japan 400 firms account for about 90 percent.

Generally speaking, the Japanese supplier industry is characterized by more "dedicated" suppliers in the sense that large Japanese suppliers are more heavily dependent on auto sales than large U.S. suppliers. In 1980, some 87.5 percent of the production value of JAPIA members was devoted to automotive products. When we examined the major U.S. suppliers, we found that of the 78 top suppliers identified, only 6 obtained more than 35 percent of their sales from the automotive manufacturers, while one-third of the 78 reported that they obtained 5 percent or less from the automotive manufacturers. One should not overinterpret these differences, however, since suppliers of raw materials are included in the U.S. data but not in the Japanese data.

This difference in the degree of dedication to the OEM automotive business by U.S. and Japanese suppliers is potentially significant in the context of the portfolio business approach adopted by many large U.S. companies. If the suppliers do not see much prospect for growth in their OEM business, as appears to be the case for many U.S. auto suppliers, we could have a "strategic mismatch" between suppliers and OEMs. This, in turn, may lead the parent company of the suppliers to be unwilling to invest significantly in improving their service to their OEM customers. Alternatively, the great diversification of the large U.S. supplier firms could lead to a carryover of their technological and other expertise into the OEM business of their auto suppliers.

The smaller number of Japanese suppliers and their greater dedication to the auto industry fit with the manufacturer practices of working closely with their suppliers in design, quality, and productivity improvement. The well-known just-in-time delivery system, based on the principle of producing parts and components only when they are needed, symbolizes these relationships. The other side of cooperation is, of course, that Japanese auto suppliers are heavily dependent on the manufacturers, and this leaves less margin for hard bargaining and independent action on their part.

The American auto manufacturers have traditionally been more highly integrated than the Japanese. It is estimated that 45 percent of a car's purchased value is provided by U.S. manufacturers and their wholly owned subsidiaries, with 55 percent being provided by suppliers. This compares to about 25 percent made in-house for the Japanese manufacturers and 75 percent purchased outside, though Japanese OEMs typically have equity holdings in their first-tier suppliers that account for much of the total business.

The equivalent of many of the first-tier "group suppliers" in Japan would be the in-house supplier divisions of the major U.S. manufacturers. This difference in organizational structure has significant implications for the different kinds of managerial integration and control strategies, pricing mechanisms, and information exchanges between the manufacturers and suppliers. It is also in these in-house divisions that we might expect to see the kind of "dedicated" supplier relationship that characterizes the Japanese situation.

It should not be assumed, however, that in-house suppliers are automatically more responsive to the needs of auto manufacturers than group or even independent suppliers. There are some indications that the Japanese producers are able to achieve very close coordination (cemented by stock ownership and close banking ties) with their formally independent suppliers. Similarly, there is information to suggest that U.S. manufacturers find that the use of independent suppliers gives them greater flexibility and responsiveness. Our research further indicates, as noted in chapter 2, that the American manufacturers seem to be moving toward reduced vertical integration for just this reason. For different reasons, the Japanese anticipate somewhat greater vertical integration in the future.

Historically, as the industry grew in the early 1960s, the Japanese automobile manufacturers organized their supplier networks and coordinated their growth consistent with separate supplier pyramids for each major producer. Since that time, some of the component manufacturers have broken away from exclusive associations, and the supply of rival manufacturers is more common, though still less common, than in the United States. Over the past twenty-five years, many of the smaller Japanese supplier firms have grown in response to the rapidly increasing automotive demand; some firms merged at the urging of the manufacturers. The outcome was more competitive production units, better able to take advantage of economies of scale.

In the Japanese industry, coordination between manufacturers and suppliers takes place, in part, through associations set up by each automobile manufacturer. Toyota, for example, has three geographically based associations of suppliers with 137, 63, and 25 suppliers, respectively. In turn, the larger suppliers will have their own supplier organizations. Although there have been changes in recent years, such forms of coordination have been lacking in the United States.

While the various differences that we have described are sometimes more apparent than real, there is no doubt that, overall, the substantive differences are significant. Currently, a major upheaval is occurring in U.S.

manufacturer-supplier relationships as the manufacturers seek to reduce costs and, in those situations they deem appropriate, to learn from the Japanese model. It is anticipated that the number of U.S. suppliers will be drastically reduced over the next five years, even as the percentage of value provided by suppliers is increased. Moreover, U.S. suppliers are increasingly facing competition from foreign suppliers (especially the Japanese) located abroad and domestically. Similarly, the whole structural relation between manufacturers and suppliers is being reevaluated in areas of scheduling, quality, investment in technology, design, early vendor selection, and purchasing decisions. Thus, we might anticipate a tempering of the differences we observed. Yet, it is unlikely that many of the distinguishing features between them will disappear completely.

Purchasing Policies and Practices

In order to understand manufacturer-supplier relations, a basic starting point is the structure, policies, and practices of the customer base. Since the purchasing function is the most direct (and, at least formally, the most important) linkage between manufacturers and suppliers, there is a need to understand the environment provided by the operations of customer purchasing departments.

Our research was based on a series of interviews with the top purchasing management of six automobile manufacturers, three each in the United States and Japan. These interviews were usually supplemented by written material, organizational charts, statistics, etc. provided by the firms interviewed.

Our first observation is that there are some areas (mainly philosophical) in which most of the firms interviewed seem to be following similar paths. There is also a wide diversity in the organization and operation of purchasing departments among the companies studied, even within one country. Thus, we were faced with distinguishing between intercompany differences and more general differences between the industries in the two countries. Finally, we would point out that purchasing organizations and activities in both countries are presently in a state of change.

First, let us summarize the similarities across systems in the two countries. All the companies interviewed espoused the fundamental commonality and interdependence of interests among themselves and their suppliers. This is meant to be the cornerstone of the purchasing department's operating philosophy. It is fair to say that American firms are only now building this interdependency notion into their operating methodology, while Japanese automakers have based much on it for twenty to thirty years. Furthermore, in both countries and among all the firms there is a recognized need to establish competitive parts pricing in order to be able to provide the automobile customer with the best value for the money in the vehicle marketplace.

Japanese firms, however, appear to have reached the conclusion that the consumer is the final arbiter of parts pricing somewhat sooner and more directly than their American counterparts. At present, the means to the ultimate end of consumer price satisfaction seems to differ significantly, particularly with regard to price setting between manufacturers and suppliers. Traditionally, American automakers have relied on direct market forces among suppliers to act as the disciplining mechanism, while the Japanese have relied on more subtle forms of indirect competition to reach the same ends. However, Japanese firms seem to be trying some elements of the American approach and vice versa. Both American and Japanese automakers are expanding their offshore sourcing, albeit for rather different reasons, as discussed in chapter 3. In the short term, at least, it would appear that this will have larger quantitative effects in the American case.

Turning now to what seem to be systematic differences between the purchasing systems in the two industries, we begin by observing that Japanese and American companies differ significantly in the apparent application of delegation-of-authority concepts within purchasing staffs, with American firms applying them more extensively (at least formally). However, it is not clear that the actual application of "delegation of authority" in decision making is as different as the formal differences suggest.

Japanese automotive firms almost always have "purchasing engineering" functions as part of their purchasing organizations, while these are almost universally absent in American firms. On the other hand, there does appear to be a significant influx, very recently, of buyers and managers with significant technical credentials in American purchasing departments. However, with very few specialized exceptions, they are in traditional line and staff roles and not serving a "purchasing engineering" function.

As was mentioned above, perhaps the most striking operational difference between purchasing in the two industries is in the area of part pricing. Japanese automakers and their suppliers have succeeded to a much greater extent than their American counterparts in putting price determination on an "intellectual" basis by using the so-called "target-cost" system. Through this system, price is directly related to cost, and the purchasing group has sufficient expertise in value engineering to understand their suppliers' costs. The advantages of this approach are being recognized in American circles, and efforts are under way to determine to what extent and in what way it should be applied there. It should be recognized, however, that once the formalities are eliminated from consideration, the reconciliation of pricing through negotiations on an informal basis are strikingly similar in the two countries.

There are a whole list of differences observed between firms in the two countries that appear to derive from the fundamental organizational forms of the firms rather than specific decisions with respect to purchasing. Thus, for example, Japanese purchasing departments are very highly centralized, following the basically centralized functional organization of Japanese automakers. On the other hand, many American automakers have decentralized purchasing functions, reflecting the decentralized nature of their operations.

The different organizational roles of divisions and group companies in the American and Japanese automotive industries is very difficult to study. For example, we found it nearly impossible to learn much about specific bases for determining transfer prices between divisions of American automakers. However, in a broader context, a comparison of "target costs" and "transfer pricing" in actual application might provide significant information relevant to motivators and demotivators to technological change, market response, etc.

Finally, we would observe that the purchasing departments in the two countries have been studying each others' approaches. In both fundamental and more detailed ways, there seems to be a sharing (certainly largely indirect) of thoughts and methods. While we would hardly expect to see the two systems identical five years from now (there are indeed good reasons for many of the differences), we would expect to see more similarity then than now, particularly in the fundamental areas of balance between interdependency and market forces and the effects of vendor selection and price determination on this balance.

Structural Characteristics of the Japanese Supplier Industry

Introduction

Past analyses of the Japanese supplier industry have been hindered because of the complex structure of the industry. The Japanese automotive-supplier industry has often been characterized as vertically organized. In other words, there exists a hierarchical, multi-tier structure among suppliers in terms of customer-supplier and equity-holding relationships. The first-tier suppliers are suppliers that have direct relationships with vehicle manufacturers; the second-tier suppliers are those that have indirect relationships with vehicle manufacturers through first-tier suppliers, and so on. Because of the complex interrelationships of this multi-tiered system, it is difficult to understand the function of suppliers at the lower end of the system. Thus, any attempt to describe the structural characteristics of the second-tier parts and components suppliers would enhance our understanding of the industry.

Based on this need, the Japanese research team conducted a survey of first- and second-tier suppliers. This, of course, is only a first step to understanding those suppliers at the lower rungs of the ladder. Moreover, suppliers of raw materials are not included in this survey.

Since the purpose of the survey is to understand the structural character-istics at this second level, the sampling of the suppliers was designed so that the relationships between the second- and first-tier suppliers would be clarified. Two major OEMs were asked to identify four major first-tier suppliers each. These eight first-tier suppliers were asked in turn to select sixty firms that they considered their second-tier suppliers, and all the quantitative data from the second-tier suppliers were compared with that from the first-tier suppliers.

However, the first-tier suppliers that were used for the sampling of the second-tier suppliers tend to be larger than the average of all the first-tier suppliers; related to this characteristic is the fact that they not only supply OEMs but also are supplied by second-tier suppliers in our sample. Therefore, our initial comparison between the first- and second-tier suppliers could have been biased in favor of these larger first-tier suppliers. In order to compensate for these differences, the two OEMs were asked again to select an additional eleven first-tier suppliers each, so that we have a total of thirty first-tier suppliers that are intended to be representative of Japanese first-tier suppliers.

In summary, the data to be reported is based on the threefold distinction we have outlined above. The first category is the first-tier suppliers (sample of eight firms). The second category is second-tier suppliers (sample of sixty firms). The third category is composed of the first category of eight first-tier suppliers plus the supplementary group of twenty-two first-tier suppliers, for a total sample of thirty first-tier suppliers. For ease of presentation, we will refer to this third group as the combined first-tier suppliers.

Characterization of Second-Tier Suppliers

Corporate Characteristics. The Japanese second-tier suppliers have, on the average, about 340 employees, roughly one-eighteenth of the employees of the first-tier suppliers and one-eighth of the combined group of first-tier suppliers. The second-tier suppliers, on the average, held capital in the amount of about $375,000 in 1981. The average capital of first-tier suppliers is roughly fifty times more than the capital of the second-tier suppliers and twenty times that of the combined first-tier group. Annual average sales for the second-tier suppliers were roughly $2.2 billion in 1981. The sales for the first-tier suppliers were about twenty-eight times more than for the second-tier suppliers and twelve times more than for the combined group. Combined with the figures concerning the number of employees and capital, the data reveals that the second-tier suppliers have lower sales per employee.

Relationship with Manufacturers. The relationship of the second-tier suppliers with the OEM is less close than that of the first-tier suppliers, in terms of equity-holding relations, reliance on the automotive market, and the recruitment pattern of managers. There are few cases in which second-tier suppliers have direct equity relationships with OEMs. In three out of the fifty-nine second-tier firms sampled, vehicle manufacturers hold some equity, while all the initial eight first-tier suppliers have equity-holding relationships with OEMs, and twenty-three out of the thirty combined first-tier suppliers have such relationships with the OEMs.

The percent of sales that are related to automotive products is lower for the second-tier suppliers than for the first-tier suppliers. The second-tier suppliers, on the average, relied on automotive products for 82 percent of total sales in 1981, while the first-tier suppliers relied on automotive products for 98 percent of their sales; the combined group of first-tier suppliers relied on auto products for 92 percent of their sales.

There is a closer degree of communication at the managerial level between the first-tier and OEMs than between the second-tier and OEMs. This is manifested in our finding that about 80 percent of the managers in the second-tier suppliers have never worked for an automobile manufacturer, compared to about 54 percent for the managers in the first-tier suppliers and 66 percent for the combined group of first-tier suppliers.

Corporate Human Resources

Employee Characteristics. In the survey, employees were divided into the following three groups: (1) production workers; (2) white-collar employees; and (3) part-time and seasonal workers. The proportion of part-time and seasonal workers for the second-tier suppliers is a relatively high 10.9 percent, compared to 4.2 percent for first-tier suppliers and 4.4 percent for the combined group. This implies that the second-tier suppliers have more flexibility in adjusting their labor force than the first-tier suppliers. In contrast to the ratio of part-time workers, the ratio of white-collar employees in the second-tier is somewhat smaller than in first-tier suppliers: 25.6 percent in contrast to 29.1 percent for the first-tier suppliers and 31.0 percent for the combined group.

With regard to education, the percent of white-collar employees with college degrees is lower for the second-tier suppliers (32.0 percent, in contrast to 39.1 percent for the first-tier suppliers and 35.1 for the combined first-tier group). However, 1.6 percent of the blue-collar employees of the second-tier suppliers hold college degrees, while there are no college graduates working as blue-collar workers in the first-tier suppliers.

With regard to age, the employees of the second-tier suppliers, on the average, are a little bit older than those of the first-tier suppliers. For blue-collar workers, the average age is 35.7, compared to 32.2 for the first-tier suppliers and 33.6 for the combined first-tier group.

Half of the second-tier suppliers who were surveyed do not have a union, whereas all the first-tier suppliers have a union. The percentage of employees who belong to the union is higher in first-tier suppliers than in second-tier suppliers (69.0 percent for blue-collar workers in the second-tier suppliers, compared to 95.3 percent in first-tier suppliers and 90.8 for the combined group).

Management Characteristics. In the survey, we also investigated managers who are members of boards of directors or above (corporate officers). As far as their functional background is concerned, the majority of managers in second-tier suppliers have backgrounds as engineers (56.1 percent of their managers have engineering backgrounds). It is interesting to note that there is no difference between the second-tier and combined first-tier group, but that the percent is lower for the first-tier suppliers (50.6 percent of their managers are engineers).

Half of the managers in the second-tier suppliers have worked for their present firm for twenty years or more, and over 70 percent have worked for their present firm for ten years or more. There is no difference between the second- and first-tier suppliers, but the seniority of the managers in the combined first-tier group is shorter (only 34.4 percent of the managers have worked for their present firm for twenty years or more).

Average Wage and Nonwage Benefit

It is a common view that working conditions are worse among the second-tier suppliers; hence, it is essential to compare the average wages and nonwage benefits of first- and second-tier suppliers. In the survey, only the wage for blue-collar workers was investigated, and our data does not include bonuses, which are quite common in the Japanese industry. The absence of bonus data may serve to minimize the actual differentiation between first- and second-tier suppliers. We also did not compare first-tier supplier wages to those of the OEMs, so readers should not make any undue assumptions in this regard.

Concerning the difference in wage between suppliers, the hourly wage for employees in second-tier suppliers is about 95 percent of that in first-tier suppliers and 87 percent of that in the combined first-tier group. To put this matter in a broader context, we can examine the differential cash earnings (contractual wages plus special cash payments including overtime and bonuses) by firm size in the transportation-equipment sector (the automobile industry makes up roughly 71 percent of the total employees in this sector). Based on data from the Monthly Labor Survey carried out by the Ministry of Labor, we find that employees in firms of 100-499 employees make 82 percent of those in firms 500 or larger; employees in firms of 30-99 employees make 67 percent of those in firms of 500 or larger. Examining production workers and white-collar employees separately yields roughly similar statistics. To be sure, we might assume that if we had examined only automotive industry employees, the wage gap might not be quite as great.

These findings are not inconsistent with our survey findings: the average size of second-tier suppliers in our sample was 340 employees, with none below 100 employees. The Ministry of Labor data do make clear that once one gets down into the third- and fourth-tier firms, the wage differential grows significantly. According to the Japanese Census of Manufacturers, over 20 percent of auto-industry employees were employed in firms of 100 or fewer employees. Thus, the competitive advantages associated with these wage gaps cannot be overlooked.

As far as nonwage benefits are concerned, the employees at both types of suppliers have, in general, access to a majority of the benefits, although employees in the first-tier suppliers have more access. The benefit that exhibits the greatest difference between both types of suppliers is "subsidy for housing." It is offered by almost all the first-tier suppliers surveyed, whereas only forty-two out of the fifty-nine second-tier suppliers offer this benefit. Again, this does not seem to constitute a major difference.

In order to see approximately how high the cost resulting from nonwage benefits might be, suppliers were surveyed about their average cost index. The present average wage of each firm then was indexed at 100 for reference. In the Japanese industry, nonwage benefits are classified as obligatory (legally required) and voluntary. Obligatory fringe benefits are, for example, medical insurance, pensions, employment insurance, retirement benefits, and benefits in case of disability.

The first-tier suppliers exceed the second-tier suppliers in both obligatory and voluntary fringe benefits. However, the difference in obligatory fringe benefits between both types of suppliers is smaller. The cost index for obligatory fringe benefits is 13.5 in second-tier suppliers, compared to 15.3 in first-tier suppliers and 15.3 for the combined first-tier group. The index for voluntary fringe benefits for second-tier suppliers is 8.5, compared to 15.8 in first-tier suppliers and 10.8 for the combined group. Even though all these differences between first- and second-tier suppliers might seem less than one might have expected, they may still operate to provide an important competitive edge. Moreover, the competitive edge might be still larger as one gets into third- and fourth-tier suppliers.

Corporate Strategy

Diversification of Business. The majority of the Japanese second-tier suppliers have already diversified their business (though the degree of diversification is not reported). Among the sectors into which the second-tier suppliers have diversified, nontransportation manufacturing is the major one, followed by nonautomotive transportation manufacturing. Yet, compared to the figures for the first-tier suppliers, the proportion of second-tier suppliers who have diversified into the automobile aftermarket is low.

Product Development and Engineering

It is interesting to examine the relationships between second-tier suppliers and their customers with respect to specification design. The suppliers were asked to list their major products and to indicate which are produced according to specifications supplied by customers and which according to specifications designed by suppliers. Second-tier suppliers produce 76 percent of their products with specifications supplied by their customers. In contrast, the majority of products of the first-tier suppliers are produced with their own self-designed specifications.

It is rational to assume that the firm's technological capability is, to some extent, a function of R&D investment, which is estimated by noting what percent of sales the suppliers have spent for new product R&D corporate wide. The average percent of R&D investment to total sales is 0.9 percent in second-tier suppliers, whereas the corresponding figure is 2.9 percent in first-tier suppliers and 2.3 percent for the combined first-tier group.

Purchasing

The annual amount that the second-tier suppliers purchased in Japan is, on the average, about $10.4 million. The amount purchased by the first-tier suppliers is about thirty-eight times more than the second-tier suppliers and twenty-six times more for the combined first-tier group. Concerning the number of domestic firms from which the parts suppliers purchased, the second-tier suppliers have, on the average, fifty-five suppliers. The first-tier suppliers have five times more domestic suppliers, and the combined first-tier group have three times more.

Offshore sourcing remains at a low level in the supplier industry. Very few second-tier suppliers (only three out of fifty-nine) have purchased abroad, whereas eleven out of all thirty-one first-tier suppliers have engaged in offshore sourcing.

Present and Future Perspective

Historically, the Japanese parts suppliers were more or less founded by the manufacturers, and, until recently, the parent manufacturers typically gave the suppliers considerable guidance and technical support. This relationship between the suppliers and manufacturers can be called a "vertical relation." Now they are separate companies, though they are related to each other through equity holdings.

Emergent developments may be labeled a "semihorizontal relation" whereby in some areas the suppliers' technological capability overlaps with that of the manufacturers. In the future, more and more technological initiatives are expected to come from the suppliers. Through our survey, however, we found that these changes are occurring only in the first-tier suppliers; they have not reached the second-tier suppliers.

As to the ranking of important factors for maintaining or improving market share, the factor ranked first by the first-tier suppliers in the past is "customer evaluation of quality performance." At present, "customer evaluation of technical qualifications" is ranked highest; in the future, "technical initiative" is expected to be foremost. However, there is no change in the factor ranked first by the second-tier suppliers, with "quality performance" ranked first in the past, the present, and the future.

As to the ranking of the most important function in the customer's organization that accounts for market share, the function presently ranked first by the first-tier suppliers is "purchasing," but it is "product engineering" in the future. However, there is no change in the second-tier suppliers' response, with "purchasing" ranked first presently and in the future.

Finally, let us describe the differences in the future perspectives between first- and second-tier suppliers. There are many technological changes occurring in the automotive industry. Those most frequently mentioned are: robotics, CAD/CAM, and FMS. We asked the suppliers, What is the most important technology that contributes to productivity improvement? The responses from first-tier suppliers are different from those of second-tier suppliers. Both first- and second-tier suppliers ranked robotics as the most important technology. FMS is ranked second by the second-tier suppliers, while it is not regarded as important by the first-tier suppliers. Instead, they list new tooling quality as the second most important technology. We can conclude that a shift to FMS is recognized as more important by the second-tier supplier level rather than at the OEM and first-tier level.

Japanese manufacturer-supplier relationships have been characterized by stable contracts, earlier product and process involvement, single sourcing, and quality and manufacturing consciousness instead of price and product consciousness. As to the ranking of factors contributing to competitiveness, quality is ranked first and manufacturing technology second by both first- and second-tier suppliers. As far as the third position is concerned, first-tier suppliers regard "product technology" as important, while second-tier suppliers regard "price" as important. The first-tier suppliers regard "price" as least important. These findings provide some support for the conventional views, especially at the first-tier level.

To sum up, product technology is becoming important at the first-tier level, while price remains more important at the second-tier level. Nonetheless, quality and manufacturing technology continue to be the basis for competitiveness.

We also asked, What will be the important practices that serve to improve service to customers in the coming decade? Earlier product and process involvement is ranked first by first-tier suppliers but third by second-tier suppliers. Second-tier suppliers ranked "less-volatile order release" first, while the first-tier listed it second. None of the first- and second-tier suppliers regard "exclusive sourcing" as important.

In conclusion, "earlier product and process involvement" is more important in the relation between OEMs and first-tier suppliers than between first- and second-tier suppliers. To second-tier suppliers, a stable contract is more important since they are more likely to lack such a commitment. Exclusive sourcing seems less important than foreign observers commonly believe. The survey results reveal a variety of new trends in the Japanese supplier industry, but they also suggest variation in the significance of these trends for first- and second-tier supplier levels.

Technological Innovation in the U.S. Supplier Industry

The process of innovation in the auto industry cannot be understood without an appreciation of interorganizational dynamics, including manufacturer-supplier relationships. Technological expertise resides in many places throughout the industry, and organizational arrangements will ultimately determine whether this expertise is fully tapped with knowledgeable parties working cooperatively. Moreover, product innovation cannot be isolated from process innovation. Product design must be informed by manufacturing feasibility and costs to achieve a cost-effective and quality product. Organizational arrangements will influence the extent to which product and process innovation is integrated in order to achieve an optimal mix between product features, manufacturing ease, and manufacturing costs. This appears to be effectively accomplished in Japan, in part through long-term, cooperative relations between manufacturers and suppliers. As argued below, this has been less characteristic of the automotive industry in the United States, though programs are already in place that will, in a sense, shift organizational relationships closer to the Japanese model.

To understand the product-planning process in the United States, case studies of particular components were undertaken (Working Paper 15). The case studies provide an in-depth description of formal mechanisms and informal interactions for the development and ultimate manufacture of window regulators and disc-brake calipers. By selecting specific components, the investigation was provided with a logical rationale for the selection of persons to interview in the auto-manufacturing organizations. The summary included here focuses on generic problems and processes that influence technological innovation. Observations are based on interviews with persons in product planning, engineering, purchasing, scheduling, quality assurance, and assembly in two large U.S. automotive manufacturers. Visits to supplier engineering and manufacturing facilities were also undertaken for most U.S. manufacturers of disc-brake calipers and window regulators. Some of these suppliers were internal to the automotive companies, and some were external suppliers.

Burns and Stalker[36] suggest two opposite forms that an organization can take when responding to rates of technological change. A *mechanistic* system appropriate to stable conditions is characterized by the following:

specialization of organization by function
hierarchical structure and reconciliation by superiors
focus on improvement in *means* of doing tasks
emphasis on systems and procedures, rights and obligations
systems and procedures defined by functions
systems translate into job specifications
primarily vertical interactions
loyalty and obedience

By contrast, an *organic* system appropriate for changing conditions is characterized by the following:

organizational relationships by network structure
commitments to tasks and to progress
adjustment and redefinition of tasks through interaction
communication of information and advice rather than instructions and
    decisions
importance of expertise and contribution of special knowledge
continued participation and interaction with others

We observe that one organizational type is not necessarily better than the other. For stable conditions, in fact, it is probable that the mechanistic form is superior to the organic.

The product-innovation process and the other forces impinging on the U.S. industry suggest that it is undergoing a change in technology and market conditions that would make the organic form of organization superior to the mechanistic. However, many comments lead toward the conclusion by technical employees within the manufacturer's organization that the organization may be classified as mechanistic. Their comments are consistent with Burns and Stalker's observations that in mechanistic organizations persons tend to perceive the rest of the organization as part of the problem to be dealt with, rather than as a source of expertise or knowledge. For example:

A quality engineer: "We routinely challenge design engineers. . . . They usually do not have warranty responsibility or charge backs. They look for minimal design cost. It's a somewhat adversarial system."

---

[36]Tom Burns and G. M. Stalker, *The Management of Innovation* (London: Tavistock, 1961).

A quality engineer: "Quality procedures tended to split responsibility for quality. A supplier would say, 'But the manufacturer approved this set-up, so why is he blaming me for this screw-up?' "

A component engineer: "It can be difficult to work with internal supplier X. They're not afraid of us. They used to change components without notification, but they do that less now. But we are part of the same company, so the arguments can become bitter."

A component engineer: "Advanced engineering is unrealistic. It's left to us to make sure the part is cost effective and will work."

This was even more apparent in visits to external suppliers who often have quite tenuous relations with their customers, even when they have had long-term relations extending across decades. For example, in several instances we heard descriptions of supplier designs appropriated by the customer and taken to another supplier who came in with a lower bid. The complexity of the relationships among several levels of component assemblers (both independent and within the auto manufacturers), the suppliers of parts to the assemblers, and the component-specification practices among the auto manufacturers' design engineers seem largely responsible for ambiguities about the origin of innovations. For example, it is not uncommon to hear the process of developing a new component specification described in the following way: During advanced and product engineering, OEM engineers may interact with existing or potentially qualified suppliers, soliciting their ideas and critiques of design concepts as they develop. These interactions are usually with separate individual suppliers so as not to create overt release of suppliers' proprietary ideas to competitors. After all ideas are received, the manufacturer's engineers prepare a complete detailed set of specifications and drawings of the manufacturer's component requirements. The drawings are prepared as the manufacturer's drawings, so that none of the suppliers who participated in the earlier development of concepts readily identify which, if any, of the competitors contributed the ideas. Moreover, the manufacturer may send requests for quotation to suppliers who did not participate in the engineering development process. Hence, these suppliers gain access at no cost to the technical innovations provided and sometimes already patented by competitors. Although the manufacturers report they are taking steps to minimize such occurrences, they do admit that occasionally suppliers who have made contributions during design engineering may be overlooked in the process of requesting quotations, either because buyers were not aware of the suppliers' efforts, or because requests for production quotations were processed by a different purchasing group than the department that buys engineering services, testing, or prototype development.

In those instances where a supplier holds the patents on some aspect of the component awarded to an unlicensed competitor or to an internal plant, legal action is usually avoided in favor of negotiation. Generally, a regular supplier feels it is somewhat awkward to bring suit against the auto manufacturer since he may jeopardize his on-going business in other components with the same manufacturer. Furthermore, smaller suppliers either may be highly dependent on their relationships with the manufacturers or may simply not want to spend the time and resources required to sue a customer of the size of a GM or Ford. As a consequence, both suppliers and manufacturers tend to tolerate considerable interchange of engineering ideas without careful credit. For example, manufacturers will imply that design specifications are the manufacturer's, whereas, in fact, they may represent a collage of suppliers' inputs. In turn, the suppliers may quote a component design to an auto manufacturer that in reality is based on ideas obtained from a request for quotation from a competing manufacturer.

Given this reality, it is not surprising that the participants from the manufacturers' and suppliers' engineering functions perceive their roles in the technology-innovation process in different lights. The component-design engineer sees himself as the key figure in the manufacturer's component specification. It is up to him to take the concepts and packaging data for an advanced car concept and to provide, on time and within the cost requirements, a design that will perform according to the warranty and performance objectives. Given this role, he is perceived by the advanced-engineering staff and innovative suppliers (and sometimes by himself) as conservative and unwilling to take the risk of innovation if previous designs or design variations are available to suit the requirements.

The outside supplier perceives the component engineer frequently as competent, but with more concern about the packaging and test requirements and specifications than the problems of manufacturing and design interaction. In particular, the supplier sees himself as more knowledgeable about the specific technology of the component device since his firm may have established its credentials based on its long-time experience and contributions to the component development. Thus, the supplier feels that he knows the component technology best since he has worked with it over many years and with many auto manufacturers. In some cases, he feels his knowledge of the manufacturing technology required by the component is superior to that of the auto firm's component engineer, who may be geographically and organizationally remote from production considerations.

The supplier wants to help the manufacturer develop a component design that will lower the component cost but, at the same time, will seek to influence alternate design choices in favor of those that can be most favorable to his participation in the business. To this end, for example, he may suggest designs that permit him to use tooling that is already available in his firm. For this purpose, black-box designs provide suppliers with maximum flexibility. Thus, the customer-supplier relationship tends to foster standardization, and, in fact, risk taking is not often perceived as the road to success when manufacturers' engineering departments are shrinking and supplier sales may be threatened by manufacturer integration.

In comparing manufacturer-supplier relations between the United States and Japan, it is important to bear in mind differences in structural characteristics across nations. As discussed earlier, there are large differences in Japan between first-tier and second-tier suppliers. Customer equity holdings are much more common among first-tier suppliers, who are also much larger than those of the second-tier. In many respects, the first-tier, who may have the principal advantages of early sourcing and joint-development programs with customers, is most similar to internal suppliers in the United States. For example, over 90 percent of the disc-brake calipers for one large U.S. manufacturer are made by divisions within that automotive company. Internal supplier divisions operate under significantly different procedures than external suppliers. Since they are affiliated with the same company, it is easier for them to maintain close liaison with the various phases of the development and preproduction planning cycles. Consequently, it is also easier for them to work with advanced and component engineering in order to influence the design of the component so it is more effectively manufactured on their processes. Indeed, representatives from internal suppliers often sit on product-planning task forces.

External suppliers often resent internal suppliers for the perceived advantages in upstream information and favored treatment on procurement decisions. One external supplier felt that by the time he got the "bugs" out of the manufacturing system for a component, sales volume might go up, and the work might be pulled from him in favor of an internal OEM division. Internal divisions point to their higher labor costs as a disadvantage they must live with. Their main advantage comes from economies of scale due to larger investments in manufacturing technology. Once the division has made such investments, the OEM has an incentive to utilize in-house capacity as much as possible.

Overall, conflict and competition is evident among external suppliers, between external and internal suppliers, and among various organizational units within the automotive manufacturer itself. Exactly how this impacts the process of innovation could not be precisely discerned from our conversations in the industry. However, we did note at least one instance where an in-house innovation in a component was rapidly adopted and put in production, while a competing innovation in the same component by an outside U.S. supplier was never adopted in the United States (Working Paper 15). The latter design, which is used in Europe, is as inexpensive to manufacture as any design available worldwide and permits wider standardization of parts among different vehicles. Thus, this outside supplier is an example of a U.S. firm with a substantial reputation for its contributions to the technology of the component designs but which enjoys little if any share of the U.S. component market. It continues to seek an opportunity to use its technological strength to establish a permanent position in the U.S. markets comparable to its standing in European markets.

Current and Future Trends in Manufacturer-Supplier Relations

The first requirement of technical cooperation between suppliers and their customers is some sort of common (technological and marketing) planning base from which suppliers can develop both their strategies and specific technical solutions to the customers' needs. Our overall observation is that the Japanese automotive companies, both through their purchasing function and their supplier associations and through their engineering departments, tend to take a somewhat more organized approach to providing this planning base to its suppliers. In the United States, it is normally more incumbent on the suppliers to obtain this planning base by virtue of their proactive efforts with their various contacts within the customer's organization, although changes are evident.

In the United States, one example of recent proactive efforts by the automotive manufacturers to inform their supply base is a program at the Ford Motor Company that involves meetings with the chief executives of its major supplier companies.[37] One of the principal purposes of this meeting is to provide a forum for Ford management to explain to these executives the forward plans of the company so as to allow them to prepare to more suitably serve Ford's needs. In the case of more specific plans, Ford will quite often call meetings of the supplier or suppliers involved in a certain functional area to discuss the long-term needs in that area with respect to a given new product (vehicle) or general needs. The suppliers are then asked to respond specifically with proposals or suggestions. Every year Ford has issued a so-called "want list," describing briefly the various needs that have been identified by the various technical organizations within the company. A supplier may see an item on the list that interests him and contact the Supplier Research Program for more details.

Obviously, there is a competitive aspect to most of this information, and the more detailed information is certainly distributed to suppliers only on a "need-to-know" basis. Assuming that the supplier feels that it can assist Ford on this need, it enters into an arrangement with Ford for joint research and review programs. The advantage of this may be some Ford sharing of cost,

---

[37]Richard P. Hervey, F. Kodama, R. Wilson, and T. Yakushiji, *A Perspective on the Purchasing Function in U.S. and Japanese Automotive Companies*, Discussion Draft (Ann Arbor, MI: Joint U.S.-Japan Automotive Study, April 1983).

but, more importantly, systematic Ford review of the supplier's programs to make sure that they are compatible with the needs of the company.

The most important motivation for suppliers in working with their customers on technology development is that it should result in increased mutually profitable business for the two parties. In this respect, there should be a linkage between R&D expenditures and the awarding of contracts to produce the products developed.

In Japan, joint-development programs normally come after a *de facto* presourcing process wherein the supplier has been nominated to produce that part for its customer. It is interesting to present the observations of one major Japanese automotive supplier in this regard. This firm pointed out that, although it lost a fair number of design proposals to alternative technical ideas or cost barriers, it did have a "capture rate" of about 90 percent of the product proposals that finally were put into production. Thus, it felt that it was getting as reasonable a procurement reaction on its technical-development efforts as could be expected under the circumstances.

Recognizing the problems with respect to linkages between development efforts and procurement decisions, the major American automotive companies are increasingly moving toward what has been called "early sourcing." Under this type of program, key suppliers are determined during the design phase of a new-product program. The intent is to motivate "up-front" supplier involvement. Presumably, suppliers will make greater contributions to product designs if they are guaranteed a return on their investment of time and resources. Moreover, a closer coordination of design and manufacturing feasibility can be achieved.

One specialized type of technical cooperation in Japan that does not exist in the same form in the United States is what might be called "special task forces." Apparently, when one of the major automotive manufacturers has a significant new model or technical problem, suppliers' technical personnel are asked to come and live at the customer's engineering or plant facility for an extended period of time to work as part of the design and development team with the customer's personnel. This sounds like an ideal situation from some points of view—for example, in terms of integrating the entire product-development activities of suppliers and their customers. However, from the vendor point of view, this "draft" apparently often causes problems, since at times these special task forces are set up without particular regard to the impact on other supplier product-development or production programs. Interestingly enough, although one must assume that this type of program is more often established with major group suppliers, it is often incumbent on nongroup suppliers as well to provide personnel for such task forces.

Although such task forces in a formal sense do not normally exist between automotive companies and their external suppliers, many of the task forces in Japan take place with group companies that would be considered divisions in the United States. Interdivisional task forces within the American automotive companies often play a parallel role to those in Japan. In addition, although formally they may not be called task forces, even external suppliers will, in the case of a major development need, spend time with their customers. Hence, the distinction between the Japanese and American cases is even more blurred.

An interesting institution in Japan without an exact parallel in the United States is the Japan Automotive Research Institute (JARI). Among other duties, it is charged with using its extensive development and testing facilities to serve the entire automotive industry and, in particular, to assist smaller suppliers in the product testing required by their customers. The Japan Automotive Research Institute will physically and functionally test a supplier's product before the supplier submits it to a customer for testing. This cooperative approach seems to be a cost-efficient way of serving an important function currently required of suppliers in Japan and increasingly required in the United States. Given the trend toward more supplier functional testing in the United States, a similar institution may have merit there as well. (The government of Ontario has recently established an Auto Parts Centre, which should eventually serve these, as well as other, important needs.)

In summary, then, we see in both Japanese and American companies a variety of strategies but the same general goal of increasingly integrating their suppliers into the product-development cycle. The Japanese appear to be further along in this regard. However, in the survey discussed earlier, Japanese first-tier suppliers ranked earlier product and process involvement as their first priority for improving service to customers in the coming decade. Once again, we observe that, although the methods may be different, companies in both countries seem to be evolving along somewhat parallel paths and, perhaps, converging in some respects.

Chapter 10
Human Resource Development
and Labor Relations

Introduction

The objective of this chapter is to report on the major findings from our research on human resource development and industrial relations in the U.S. and Japanese automotive industries. The need for brevity in this chapter will, of course, truncate much of the description that lies behind the analysis. However, we hope to convey the areas of similarity and difference that have the most to contribute to the overall study. In particular, we seek to combine contemporary and historical perspectives on the two industries so that contextual differences (e.g., the timing of the development of the two industries) can be used to distinguish between the practices employed in each country.

The chapter is divided into three major parts. The first part describes critical areas of comparison that impinge directly on human resource development: (a) labor-management relations, (b) training and education, and (c) workshop activities. Intertwined with analyses of these areas are brief discussions of other aspects of human resource development, including employment conditions, compensation systems, and labor-allocation procedures. Emphasis is given to the first three areas in large part because of our belief that they are most broadly determining of differences between the U.S. and Japanese industries. The second part of the chapter more closely examines how attempts at innovation—technological and organizational (workshop-level participation)—are facilitated or inhibited by human resource policy and internal labor-market mechanisms. The third part of the chapter considers problems in human resource development faced by U.S. and Japanese manufacturers and offers implications for future organizational policy.

Two additional comments about this chapter's comparative perspective are in order before proceeding. First, contrary to some observers, we do not accord culture a central determining role in our attempt to account for the differences in the U.S. and Japanese systems. Instead, we assign culture a position that overlaps with and is sometimes reflected in other factors, most particularly political and economic history. Second, we suggest that it is unwise to shunt aside history in any attempt to understand the differences between the U.S. and Japanese auto industries. The two industries developed

173

in two very different historical periods. The Japanese industry began its growth in a period shaped by unprecedented economic and physical privation and under conditions not entirely of its own choosing. The U.S. industry, by contrast, was able to move very quickly into a mature stage in the postwar period, in part because of its prior history of development and the already advanced process of consolidation that had dramatically whittled down the number of domestic competitors. On the basis of these two points, we feel it is reasonable to consider functional similarities between institutions that, while embedded in different cultures, are nonetheless reflective of an overarching set of rules or constraints. In the case of economic organizations operating within capitalist market economies (e.g., auto-manufacturing enterprises), we can assume that certain kinds of rules are shared: e.g., the necessity of making sufficient profit, organizational stability and continuity over time, the probability that the organization will have to compete with other like-minded firms in an effort to realize their investments, and that an economic enterprise, like any other category of organization, must develop and sustain commitment to the organization on the part of all its members in order to accomplish the enterprise's goals. These rules or constraints are, of course, variable, but they do serve to limit the range of possible options available in meeting organizational needs. For example, a firm can adopt short-term versus long-term strategies for profit maximization, but it cannot long neglect profit without perishing. Thus, as we will argue throughout this chapter, the need to acquire, train, and efficiently deploy production workers in the U.S. and Japanese auto industries does not differ fundamentally; what does differ is the manner in which those needs are fulfilled—or, in some instances, unfulfilled. We will assume that these rules are sufficiently common to each organization as to make possible a comparison without relegating an explanation of differences to the "cultural black box."

We shall take points of functional similarity and attempt to account for their origin using historical, political, and economic variables and processes. The principle mechanism we will be concerned with in the fulfillment of these goals is the internal labor market. By this we mean two things: (1) the means by which firms recruit, train, compensate, and allocate labor—in our part of the study referring almost exclusively to blue-collar, production labor—and (2) the devices firms create to enable them to adapt to changes in their own organization and their environment. Given that firms in both countries have undertaken to staff and develop their organizations via internal labor markets, these mechanisms constitute the broad basis for comparison.

We should note that using the internal labor market as a model for comparison does not mean that both industries have relied on it to the same extent. This is most visibly the case in terms of the highly variable levels of employment and rates of labor turnover in U.S. firms historically. The American companies have been willing to rely more heavily on the availability of an external labor market for unskilled and semiskilled workers to fill vacancies created by high levels of labor turnover (relative to the Japanese). This has implications for job design, training needs, use of technology, and unemployment costs.

Human Resource Development
and the Blue-Collar Labor Force

The creation of an internal labor market has proven a central facet of industrial human resource development practice in both the U.S. and Japanese auto industries. In theory, an internal labor market is intended to be a pool of labor within an enterprise that can be drawn upon to fill vacancies or newly created job functions. By selecting applicants from within to fill vacancies or new jobs, the enterprise can utilize the investments it or its employees have made in general and specific training. With an internal labor market, companies can avoid reliance upon what may be highly variable external markets for skilled or experienced labor and, at the same time, provide more stable work for those they already employ. Beyond simply assuring a stable labor force, internal labor markets have the potential to provide a "skillable" labor force internal to the enterprise, thus reducing the costly and often uncertain practice of trying to acquire labor externally.

Though internal labor markets have proven central features of the industrial human resource policies of the U.S. and Japanese auto manufacturers, the way those mechanisms were constructed historically and their relative strengths in the present period are quite different. In order to draw the most from a comparison of the two cases, we focus on the following facets of human resource policy.

Labor-Management Relations

Surface comparisons of labor-management relations in the two industries usually net such dramatic appraisals as "adversarialism versus cooperation." Such characterizations are unproductive for several reasons. First, adversarialism characterized the relationship between unions and management in both countries (the 1930s and 1940s in the U.S. and the late 1940s and 1950s in Japan). Yet, as the mechanics of collective bargaining and other forms of negotiation were established in later years, U.S. and Japanese unions and management narrowed the field of contest to legally prescribed limits. Second, cooperation has not been the sole property of Japanese unions and management. A framework for cooperation evolved in the U.S. through the benefits provided union members by the health of the U.S. industry. Quite simply, the economic health of the industry provided, with the rise of unionism, a foundation for the privileged economic position of U.S. auto workers relative to employees in other sectors of the economy. Similarly, the stability of the U.S. auto companies as principal suppliers of transportation to an auto-dependent public (up until the 1970s) resulted in a generally stable demand for labor, even if fluctuating on a cyclical basis, and, equally important, a demand for (and rewards to) a stable core in the labor force. In other words, high wages, extensive benefits, stable jobs, and a legalistic handling of employment contracts established an implicit cooperation between labor and management.

Third, conflicts can potentially arise in labor-management relations in the Japanese auto companies. A disparity in views between management and labor on a variety of issues—including wages, fringe benefits, job assignments, distribution of work loads, and rationalization for productivity improvements and/or promotion schemes—can increase frustrations and innate conflicts or trigger open disputes. Recent investigations by some Japanese observers have indicated that, while the shop floor is not a seething bed of resistance, the image of calm cooperation (or passive submission on the part of workers and unions) is far from what some might think.

Where the two systems differ most significantly, however, has been in the area of communication and information sharing. Though collective bargaining has traditionally been a form of information sharing, and some companies in the United States have significantly increased the flow of information to unions about production and planning procedures, Japanese union officials at many levels have had access to similar kinds of information for some time. That access has enabled them, for example, to help determine the shape and impact of new production technologies in order to affect the balance between jobs and productivity and between returns to capital and returns to workers. Far from denying information to unions, Japanese employers have encouraged union involvement as a means by which to further integrate the purposes of the two organizations. By contrast, in America, the overall pattern has been one in which the companies have tended to distrust unions, and that distrust has been returned in full measure. Past refusals to "open the books" diminished the full potential for cooperation and hampered the employment of union experts in organizational problem solving. Some firms in the supplier industry and a handfull of individual facilities under the OEMs managed to retain a greater measure of openness through conscious action, but meaningful and continuous communication between the majority of U.S. companies and union representatives suffered from the often painful strictures of company policy and union animosity. The result has been a further encrustation of outdated work rules and a high degree of organizational inflexibility. A case in point can be found in the controversy surrounding work rules and job classifications in the U.S. industry. Many job classifications were originally developed from management efforts to create job ladders and reap the benefits of specialization. They became a part of the collective-bargaining process as unions sought to provide the measure of job security between jobs that they lacked in the company as a whole. Today, these past actions by management and labor remind us of the costs of not providing workers greater employment security and the restrictions these work rules often impose on organizational flexibility.

In the Japanese case, by contrast, high levels of communication and information sharing have enabled workers' representatives to actively join in the decision-making process. While there is some skepticism about the extent to which unions actually influence basic business decisions, it is clear (as our case studies will show later in the chapter) that unions insist on being convinced that decisions are made and implemented fairly.

However, in making the comparison of labor-management relations in the two industries, two factors should not be forgotten: (1) labor peace in the Japanese industry has been quite directly aided by the tide of prosperity that carried the industry throughout most of the 1960s and 1970s, and (2) the present accords between unions and management were preceeded by a period of intense strife in the 1950s that threatened the vitality of Japan's capitalist economy. Industrial growth, nurtured by governmental action and a stunning defeat of militant unionism in the 1950s, made possible the construction of a relatively flexible and expansive era of labor peace in the ensuing decades. Growth and prosperity can cover up many "structural warts"—as has proven the case in the U.S. until recently—but reduced growth may require adjustments on the part of Japanese companies that can kindle new fires of unrest between unions and management.

Training and Education

An organization's approach to training its employees—how it trains and how much training is available—reflects not only the availability of external sources of education (e.g., in the system of public education) but also its own commitment to enriching the human resources it employs. The impetus to forming extensive training programs in Japanese auto companies can be traced to four factors: (a) the availability of large numbers of young but unskilled workers, (b) the mandate companies accepted to develop a high-quality machinery and equipment building and repair capacity, (c) the overwhelming emphasis on product quality as a spur to product competitiveness, and (d) the *de facto* extension of guarantees of permanent employment to the bulk of the production labor force. Companies needed to invest heavily in their human capital in order to make the most of their labor, and, conversely, the permanent attachment of workers to the firm argued that they would only contribute to greater organizational productivity if they, individually, could be made more productive.

Beyond purely economic reasoning, however, training and education combine with aspects of the compensation system to more firmly bind workers to the organization. On the one hand, offers to train workers in a variety of skills make it possible for employees to expect that, over time, they may achieve greater pay, skill, and responsibility. Regular promotion ladders for those who seek additional technical education reinforce the importance of such investments. On the other hand, a considerable portion of the training and education curriculum is intended not so much to impart technical skills as it is to give the employee a sense of membership in the organization. Orientation courses for new employees, leadership classes for potential team leders and supervisors, and group discussions between supervisors and workers are all embellished with a statement about the individual's role in the overall effort. In the case of workshop activities and quality control (covered more extensively below), the effort is to establish and enhance a corporate image to which all can aspire and contribute.

That such extensive training may not be directly cost effective is not of immediate concern to Japanese employers because, beyond integrating the individual with the organization, it acts to erode a foundation for work-group cohesion in opposition to management. Two central factors are at work here. First, the offer of rewards for individual initiative breaks down the potential horizontal organization of workers according to job or, more broadly, class by establishing and assigning differential status to workers of varying skill levels and responsibilities. It attaches a value to initiative and, simultaneously, places great emphasis on distinctions between layers of workers in the same organization. Rather than constantly treating all workers as a class, therefore enhancing a "we-they" philosophy, firms rank workers within a hierarchy of all employees. Second, the practice of extending a measure of authority over the work process to subforemen, team leaders, and supervisors—who are all experienced workers—combines the broadening of promotional opportunities with the enhancement of authority via respect accorded to the older, skilled workers who fill these positions. Supervisors are respected because they know the work, not because they have been certified by a college or the company as "management material." Thus, there is less potential for animosity to develop between blue-collar and white-collar employees.

Until recently, investment in employee training and education in U.S. auto companies followed a pattern similar to that found in other aspects of manufacturing. During periods of response to market recession and intense attempts at productivity improvement, training programs were caught in a vise: on the one hand, demands for cost cutting generally reduced the softest tissue (that which seemed to contribute least to profitability) such as training; on the other hand, however, demands for training to upgrade the skills of workers who must handle new equipment and processes increased. As a result, and despite the continuation of many programs on paper, training and education internal to the organization tended to be cyclical and temporary.

One recent and encouraging sign, however, can be found in 1982 re-
negotiations of the major auto contracts: despite sharp cutbacks in employment, the companies and the union developed a plan for regular company
contributions to the development of national training and education
programs. These programs have constructed training curricula and are
offering courses to assist displaced workers in finding new jobs. The programs
are now moving towards an integrated training program for *all* hourly
employees.

One area of immediate need (which is itself enjoying the spotlight again as
the industry retools for the future) is training for members of the skilled
trades, most particularly electricians, machine repairers, and others who must
maintain equipment and facilities. That this component of the labor force
should receive focused attention is not disputed; however, it has been
longstanding practice to make the trades the sole receivers of skill-
development programs. There are a variety of reasons for this situation: (1)
members of the skilled trades are among the most crucial to the functioning of
equipment, and it is their skill that is most likely to be impacted by new
process technology; (2) sustaining a well-trained maintenance work force is a
reasonable alternative to subcontracting repair services at an exorbitant rate;
(3) incremental training of in-house skilled labor is necessary to protect the
high level of investment made in those workers who have already apprenticed
with the company; and (4) the organizational clout of the skilled trades (in the
company and in the union) makes any decision to purchase maintenance
labor risky at best. This latter point—the political power of the skilled
trades—carries over to the relative undertraining of semiskilled operatives.
To the extent that companies seek to generalize the availability of training to
all hourly employees (e.g., through training for aspects of maintaining their
own machines), some segments of the skilled trades equate such action with
outsourcing—in this case, diminishing the skills and reducing the amount of
work available to the trades.

By largely focusing on the financial side of training and enduring the
constraints on who should be trained by the strength of skilled labor, U.S.
employers have tended to underestimate the organizational value of training
and education. In the process, the responsibility for these functions has been
shifted to the public sector. The recent developments in collective bargaining
described above could, however, signal a change in training philosophy. It is
important for the future prosperity of the U.S. auto industry that these new
initiatives be sustained and expanded throughout the industry.

Workshop Activities

In the broadest sense, workshop activities include the range of interactions that occur between and among workers, management, and their
representatives. For our purposes here, however, we narrow that somewhat

to focus more on the kind and quality of communication that takes place between workers and their supervisors.

Until quite recently, workshop relations between salaried and hourly employees in American factories have been constrained by the formality of the union contract and the attempt by both sides to protect what they perceived as divergent interests. Supervisors—particularly first-level production supervisors—have borne the brunt of pressure from above to use whatever means available to meet or surpass production quotas. Their performance evaluations have had little else as criteria for success. Hourly workers have responded to those various pressures by a number of means, mostly characterized by protecting themselves from demands for increased effort. Even in light of the classic experiments of the Western Electric research decades before, which highlighted the importance of small-group activity, programmatic attempts to remove the barriers between salaried and hourly employees have tended to break up on the rocks of mutual distrust and the "bottom-line" accounting of the company. This is ironic in view of the fact that many workers, given the chance, have shown themselves eager to contribute to organization improvement in quality and efficiency.

Relative to the rapid spread of worker-participation activities in such countries as Sweden and Japan in the 1970s, there was relatively modest and very uneven progress made in the decade after the initial 1973 collective-bargaining agreement between General Motors and the UAW to establish Quality of Work Life activities. This testifies to the many obstacles standing in the way of success in U.S. industry. Notwithstanding, efforts to consciously and systematically integrate workshop activities with human resource development has achieved new momentum in recent years. Our research brought to light the long-standing efforts and successes of a number of company and union personnel to create workable lines of communication and forums for information sharing, but these tended to be scattered across the larger companies—usually the product of chance combinations of the right people under the right circumstances—or isolated in smaller supplier firms whose willingness to experiment has been bolstered by a secure market position. In the majority of companies and facilities, programs of worker participation in decision making or information sharing are just getting off the ground. (We will describe one such case in the section on case studies.) Though these new programs of workshop activity take various forms and names (e.g., Quality of Work Life or Employee Involvement), their intent is similar: creating durable means by which to encourage the involvement of workers in the direction and accomplishment of production tasks while breaking down the social and organizational barriers between salaried and hourly employees. If dollars alone are evidence that the new programs are more than just recycled versions of prior human relations efforts, then the millions being spent on the construction of training systems and the release time for employees to attend them are evidence of change.

If the programs of workshop activity in the U.S. industry bear some similarity to programs in Japan, it is no simple coincidence. In the early 1970s, QWL efforts at General Motors owed much to the Western European models, especially Sweden, on the union side and the perspective on organizational development of Rensis Likert on the management side. These earlier models were replaced, to a significant extent at least for management, in the late 1970s by the Japanese model. The Japanese approach to participation came to be seen by many as part of the "package" leading to Japanese competitive success. Yet, as has been stressed by other observers, workshop activity in Japanese firms is, at this point at least, more thoroughly integrated into the general philosophy of human resource development—both in terms of a positive practice of communication and skill development *and* as a means by which to discipline the labor force. Quality-control (QC) circle activity is the single most important device in accomplishing these ends. Quality-control circles are essential to organizational performance far beyond their apparent purpose (i.e., sustaining consistent and high levels of product quality). They provide a forum in which workers are socialized into the company philosophy. They encourage a sense of individual commitment and responsibility to an intervening layer of organization (i.e., the group) between the isolated worker and the larger enterprise. They supplement on-the-job training by bringing workers and supervisors with differing levels of expertise together to share knowledge. The circles also stimulate a form of competition within the ranks of production workers to achieve recognition within the plant, the division, and/or the company via project contests.

Participation in workshop group activities is important not only in learning skills, contributing to quality or productivity improvements, and getting to know colleagues and company organization, but also as an aspect significant of career development. Indeed, an individual's performance in such group activities is taken into account in evaluating his or her prospects for promotion. In particular, whether or not and in what way a worker took initiative as a group leader is carefully watched and evaluated by personnel managers. In this sense, although QC circle activities are in principle voluntary and informal, and pecuniary compensation for such activities is nominal, QC circle activities do play an integral role in one's career formation and promotion opportunities.

Case Studies of Innovation

In order to compare the ways in which human resource development practices and internal labor markets affect the process of organizational change, we undertook a small set of case studies in the areas of technological or process innovation and organizational innovation. The concept of innovation, we felt, not only would shed light on alterations of past practice but would also offer an opportunity to compare how human resources are used or misused in the process of change. In brief, we found remarkable similarities in outcomes from innovation in the U.S. and Japanese companies. Differences arose, however, in the circumstances promoting and conditioning the changes. That is, particularly in the case of technological changes (e.g., the introduction of computer-driven machine tools), adjustments in the number of workers needed and the reallocation of individual workers required considerable consultation between managers and workers. In the Japanese case, such discussions and negotiations predated the actual physical conversion of the specific facility by a long period of time. Union representatives, managers, and engineers continuously consulted so that there would be no surprises and, equally important, so that the workers affected could be reallocated with a minimum of personal sacrifice or cost to the company. In the U.S. case, consultation was equally intense but was spurred less by a conscious and thorough procedure for adjustment than by a realization that intransigence on either side might result in the loss of the product line altogether. Thus, in each case, we found that communication and planning helped each facility best utilize the existing pool of manpower and make adjustments in its use, but with differing degrees of ease and prior experience.

Process Innovation: the Japanese Case

In the various plants of Company C (a large auto-parts supplier) process innovations had been pursued vigorously since the company's inception nearly two decades earlier. In the period of 1975-82, significant investments were made to facilitate product diversification and to introduce labor-saving equipment in the form of numerical control units and robots. Many of these investments were also designed to bring the company in line with the production and quality-control requirements of its principal customer, a large OEM. To be more specific, the point was to introduce and/or strengthen elements of multi-station coverage by individual workers, to automate finishing steps in the production process, and to realign the organization of the production line.

Notable was the fact that, in each step of these process innovations, the number of workers required per production line was reduced dramatically, for example, from six persons operating single-function, manually operated machines in 1965-67 to one person presently controlling index and transfer machines doing comparable work. The degree of automation in this period also increased. Thus, these labor-saving process innovations created a problem of how to reallocate workers—even though the total volume of production elsewhere in the plant and the company increased to absorb the displaced workers.

The adjustment process was eased considerably by two factors: (1) the expansion of product volume and diversity in the company meant that jobs, not workers, would be eliminated; and (2) prior formal and informal agreements between company and union provided a means by which to accomplish an orderly transfer of labor between jobs and, in some instances, production facilities. Company C, like other companies in the industry, was constrained by the practice of permanent employment for regular production workers and, therefore, had the weight of that practice to uphold. Mechanisms allowing for both temporary and permanent transfers were already in place but required full prior consultation with the union to insure their fair and efficient use. In this instance, as in other cases of permanent transfer, the plan was handled carefully. The union was informed about the plan after its broad contours were set by management in company headquarters. After consulting in detail with the union, management officially informed its plant staff; the union did likewise with the local union apparatus. These groups, in turn, discussed the specific conditions of the transfer plan and transmitted that information to affected production managers and supervisors. Supervisors and workshop union representatives were consulted as a preliminary list of those workers to be transferred was devised. Union workshop officers were then prepared to hear any grievances that the workers had.

Process Innovation: the U.S. Case

In Company G, process innovation was investigated within a parts-production facility that supplied finished components for the parent corporation. In this facility, a combination of changes were being introduced: some involved the introduction of labor-saving new machining equipment and transfer devices, and others involved rather substantial reorganization of assembly operations. Both sets of changes were designed to increase the overall productivity of the facility and to aid it in competing with other outside suppliers of those components. The plant under investigation had been selected for investment after an intensive cost analysis of its competitive capacity and after serious contract bargaining with the union to achieve wage concessions and greater flexibility in work rules. These changes were instigated primarily in response to the economic troubles of the company and industry.

Nonetheless, the introduction of new equipment and process technologies threw into relief the old practices that had effectively curbed innovation in the past. Two issues were crucial. The first involved the issue of altering the system of job classifications to fit the proposed new system of work organization. From management's point of view, the elimination of some jobs and the recombination of formerly separate tasks into new single-person operations demanded that prior agreements about the content of particular jobs be modified or thrown out altogether. In particular, plans for change called for merging of the separate classifications of machine operator and job setter. Quite directly, the tasks could be combined with a modest increase in training for machine operators, and, consequently, the higher-paying classification of job setter could be eliminated. The union's opening argument, however, was that the increased skill (or training) required of machine operators could neither negate the pay differential nor the classification. The situation might have developed into an acrimonious stand-off were it not for the tenuous economic position of the facility as a whole. But, by virtue of these conditions, a compromise was reached well in advance of the changes, and both company and union had sufficient lead time to "sell" the plan to the workers who would be affected.

The second issue crucial to adjusting to the process change involved the adjustment of workers across the in-plant labor pool. Contrary to the Japanese case, some workers became unemployed as a result of the changes. Given the organizationally isolated position of this facility, workers could not be transferred to other plants because their seniority was tied to the specific plant. Thus, unlike the Japanese example—or the example of more integrated American facilities where seniority rights are extended across two or more plants in a given division or complex—some workers were laid off. The process of altering production arrangements in a specific segment of the plant thus had a ripple effect throughout the entire facility: seniority workers "bumping" (displacing) their junior coworkers all the way down the line until those at the bottom of the list were placed on indefinite layoff. This displacement process, in itself, must be recognized as a source of disruption and ineffective organizational use of manpower, even though it must also be recognized as a fundamental union strategy for maintaining and protecting equity among workers under current constraints imposed by management.

In both the Japanese and U.S. cases, process innovations were facilitated by joint consultations between unions and management and by the existence of administrative internal labor-market mechanisms for balancing jobs and employment security. However, what had transpired in the Japanese case was, in a sense, prepared for by a process of anticipating the need for adjustment, while the U.S. case was precipitated by a general environment of crisis.

Organizational Innovation: the Japanese Case

Company C developed its corporate organization from its origin in the merger of two smaller companies in the mid-1960s. In successive phases of corporate reorganization, Company C sought to create uniform business and management practices out of a history of very different techniques and products in its two predecessors. Encountering a thicket of different practices adhering to its different product lines, the company evolved a decentralized administrative apparatus for its separate plants, ultimately adjusting its policies to even out the resources available to each of its profit centers. In doing so, Company C embarked on a program of productivity increases, inventory reduction, and just-in-time production scheduling.

This latter undertaking, roughly coinciding with the first oil crisis but extending to the present period, was characterized by the introduction of the *kanban* system. While the *kanban* system is described elsewhere in this project report, it is important to note here that Company C had to deal with the implications for a broad range of organizational activities, including its production system, information-processing and management strategies, and its overall program of quality control. With the tutelage of its major customer, Company C engaged in a learning process that involved all its major departments. After extensive education for managers and supervisors in the engineering, accounting, and quality control aspects of *kanban*, the training process was extended to include the majority of production employees.

Because *kanban* impacted large segments of the labor force, its introduction alongside more uniform and extensive quality-control practices enabled Company C to enrich workshop activities. Three comments may help to underscore the importance of these changes: (1) This period of organizational innovation followed receipt of the Deming Prize (in the early 1970s) for its high level of quality control and culminated in the award of the JQC prize for maintaining that high level of quality in the intervening five years. Thus, the innovations described above were undertaken after the company had been recognized for excellence in the field of quality control. (2) It should be emphasized that the process of organizational innovation involved multidimensional improvements of the complex and interrelated subsystems. In particular, QC circle activities provided multiple and on-going opportunities for these achievements to be implemented and for workers to think, learn, and participate in the organizational innovation. (3) The labor union's support and cooperation were also an integral element of this process. It was emphasized by the personnel manager of the company that the entire organizational innovation was realized on the basis of strong and continuous union support.

Organizational Innovation: the U.S. Case

On the U.S. side, we undertook case studies of organizational efforts to innovate in their human resource development strategies by way of employee-participation programs in several firms. Over the course of the project, we interviewed over ninety people involved with one or another aspect of employee participation in two major OEMs. This included representatives from top corporate positions, international union leadership, divisional salaried managers, and plant personnel. The bulk of those interviews (some seventy-five altogether) were conducted in a parts-division plant attached to one of the major OEMs. This latter group included a cross-section of the plant population, ranging from the plant manager and his operating committee, representatives from the major production and staff positions in the plant, union representatives, the various layers of production supervision, and randomly selected hourly employees.

The employee-participation program in this plant began in 1980 and has been augmented in recent years. Two labor-management committees co-ordinate its various activities and plan for future growth. To date, the most visible aspect of the program for employees has been the construction of over a dozen problem-solving groups (functionally the same as quality-control circles). These problem-solving groups consist of workers drawn from specific departments or zones in the plant and tend to be composed of eight to fifteen members—usually all but a few are hourly employees. In some departments of the plant, however, it is mandatory for salaried supervisors (especially first-level production supervisors and/or general supervisors) to participate. After a period of training in group decision making and problem solving, the groups meet on a regular basis: usually once a week before or after work, with pay. At present, however, less than 10 percent of the plant population is actively involved in one or another problem-solving group. Plans were being made to increase the number of groups and to extend the training to workers who might not join groups but who are eager to have the training.

Our investigation netted a number of important findings and, as might be expected, questions, as well. There is no question that efforts to increase the involvement of hourly employees in the organization of work have been met with very high levels of enthusiasm. The overwhelming majority of workers we interviewed felt strongly that the effort was long overdue; this included those who were actively involved and those who had yet to join. For those workers who are willing to temporarily suspend judgment about underlying motives, the feeling is a real one that for once, as one assembler told us, "somebody is finally listening." Other evidence, however, is more difficult to come by and, when available, tough to decipher. We were told that in this facility, grievance and discipline rates had dropped since the inception of the participation program and that quality ratings had been strengthened. Given the relatively small percentage of workers involved and the general decline in problems in labor relations, emanating from the hard economic times, the relationship between participation programs and improved management-labor relations is difficult to assess. Yet, there is considerable room for optimism.

However, the problems these programs face are manifold and deserve mention. They require suspension of past hostilities and prejudices, as well as a willingness on the part of management to forego the cost-accounting mentality that sunk previous efforts. They demand freedom from the rigid organizational politics on the part of unions—an especially difficult situation given the changes union officials must cope with in the face of reduced membership and concerns with job security and lost wages. They can only survive if they are accompanied by structural changes—not simply convenient shifts in attitudes on the part of management—that will affect the protected positions of unionized workers and the past claims to arbitrary authority on the part of upper- and middle-level managers.

Conclusions and Policy Implications

With the emergence of adverse economic conditions for the U.S. auto industry in recent years, dramatized to some extent by increased imports from foreign competitors, the limits of conventional human resource development practice and industrial relations have been recognized increasingly by management and labor. The lack of flexibility in conventional practice has been keenly felt. Various attempts are now being made in many plants across the country to gain flexibility in issues of human resource allocation, productivity, and product quality. Unions have accepted wage moderation and changes in work rules, while demanding a greater degree of employment security in the face of deteriorating employment opportunities for the industry. While such attempts as Quality of Work Life campaigns, Employee Involvement programs, QC circles, and joint consultation are still maturing, measures to enhance flexibility and information sharing will prove to be even more indispensable in the future. For the survival of the industry and the economic security of workers in the face of ever accelerating technological change, reform in the American structure is a high priority.

One of the areas in which the need for reform is most evident is in the organizational practice of decision making and worker participation. While the concept of participative decision making is not a cure-all for the U.S. auto industry, it can go a long way to establishing a more open and flexible relationship between workers and managers. Little can be changed, however, unless steadfast efforts are made on the part of the unions and the companies to introduce structural changes alongside efforts to change attitudes. The survival of participation programs, in fact, depends on structural change. Four interrelated factors must be addressed: (1) moving involvement beyond those who are already "quality conscious," (2) extending the pitch of the interest curve by increasing the breadth and depth of training and the terrains of investigation for the problem-solving and quality-control groups, (3) developing a combination of incentive and requirement for management groups to respond to and consult with hourly workers in the process of their own decision making, and (4) constructing a training and incentive program for lower levels of production supervision that rewards supervisors and higher levels of management for their participation in problem-solving groups and their own inculcation of an environment that encourages communication and flexibility between themselves and hourly workers.

In Japan, too, the basic background factors that supported the vigorous performance of the auto industry have been changing substantially in recent years. The labor force is aging rapidly, and the prospects for growth in the market are diminishing. The need for internationalization through direct foreign investment is growing, and new waves of technological innovation are threatening employment while suggesting new frontiers. In short, the conventional structure of Japanese management is now faced with a serious challenge, particularly whether management can effectively adapt to current and foreseeable major structural changes.

One area of pressing concern, from the point of view of human resource development, resides in the promotion system. The promotion system is facing problems arising from the slower growth of companies relative to a decade ago and from the aging of the labor force. The slowing down of the speed of promotion and the resultant decline in morale and incentive weigh heavily on the traditional sources of motivation for Japanese workers. The remedies that companies have been trying include enrichment of rank qualifications, and these are directly related to wage ranks and other fringe benefits. However, the depressing effect arising from the slower speed (or lowered probability) of promotion in the line hierarchy does not seem to be offset by the utilization of such systems of reward ranking. Now the challenge will reside in how work organizations can be rearranged so that workers can still motivate themselves and in how opportunities for promotion can be equitably distributed in the face of an increasingly narrow range of promotion possibilities.

Under the circumstances outlined above, a management strategy that will be the most important for the U.S. and Japanese auto industries is systematic and effective human resource development. Fostering a productive, creative, and adaptable work force will be the single most important measure and will be even more so the cornerstone to facilitating further development of the industries in the forthcoming era of rapid technological and structural changes.

Another important element that needs to be underlined is the flexibility in attitude, structure, and policy of both management and unions. One of the crucial conditions that determines flexibility is the extent to which both parties understand and share information about the situation. Cooperative efforts toward joint problem solving will need to be pursued on both sides. It is fair to say that the Japanese auto industry has deeper experience in this regard and that the commitment of the U.S. auto industry to pursue a systematic effort in human resource development for production workers in the absence of economic crisis remains problematic.

Chapter 11
Human Resource Development:
Management and Technical Personnel

Introduction

In the discussion about the current competitive position of the U.S. and Japanese automotive industries and their prospects for the future, great attention is given to issues related to blue-collar workers, their productivity, wages, and labor-management relations in general. The role of white-collar workers has so far received little emphasis.[38] Yet, not only do the white-collar employees comprise today nearly one-third of the total labor force in the automotive industries of both countries, their share is expected to grow with the changing structure and content of jobs brought about by technological change. Moreover, we saw in chapter 7 that their contribution to the MCD is substantial. In this environment, an effective management of white-collar human resources becomes one of the key conditions for long-term success in the world's auto markets.

For example, within the engineering function, which constitutes probably the most critical area of white-collar employment in the auto industry, increased technological sophistication and the complexity of both the final product and the manufacturing process will require large numbers of qualified technical employees to maintain and improve the competitive position of each vehicle manufacturer, as well as their suppliers. Recruiting, developing, and motivating this group of employees is already one of the critical tasks facing corporate human resource management professionals.

In this context, the objective of this chapter is to present a review of current conditions of white-collar employment in the U.S. and Japanese auto industries, with a particular emphasis on the corporate human resource management policies. The chapter begins with an analysis of the composition and demographic profiles of the white-collar labor force. This is followed by a discussion of the corporate structure, the role of the personnel function, and specific human resource management functions targeted at white-collar employees. It concludes with a summary of current trends and their implications for human resource management policies.

---

[38]Within the scope of this chapter, white-collar employees are defined as managers, supervisors, administrators, technical professionals and support personnel, sales employees, and clerical workers.

The Composition of the White-Collar Labor Force

The proportion of the white-collar labor force in the total labor force is similar for automotive firms in both countries, ranging from 26 to 33 percent. In fact, it seems that variations within each country are sharper than the difference in the "average" for the United States and Japan. For example, in American Company A, the ratio of white-collar employees relative to the total labor force is 26 percent, while in Company B this proportion reaches 33 percent. In Japan, the white-collar share for Company X is 29 percent, for Company Y, 33 percent. Among parts suppliers in both countries, the proportion of white-collar labor is slightly smaller, 24 to 29 percent depending on size. Some of the differences between firms can be accounted for by differences in job classifications at the interface between white-collar and blue-collar jobs. However, especially in the case of the U.S. firms, differences in "layoff" policies in the period of retrenchment also may play a role, as pointed out below.

In the U.S. firms, 68 to 72 percent of the white-collar labor force are salaried, exempt employees. Among these exempt employees, about 32 to 37 percent are placed in supervisory and managerial positions, accounting for approximately 10 percent of the total labor force. However, only about one-tenth of the supervisors and managers are eligible for an incentive bonus, a distinction of considerable status. In Japan, these ratios are again similar, although direct comparisons are difficult not only due to differences in job classification but also due to the exclusion of female workers from the available personnel statistics. It should be kept in mind that the term "salaried, exempt employee" has little meaning in Japanese firms where both blue-collar and white-collar employees are paid on a salaried basis and where lower-level white-collar personnel belong to the same union as blue-collar workers in the same firm.

In Japanese Company Z, supervisors and managers (including employees of equal status without managerial responsibilities) account for 8.2 percent of the total labor force. In Company X, the number of managers (excluding first-level supervisors) equals just over 3 percent of the total employment, or 13 percent of white-collar employees. About 25 percent of these are ranked as assistant general manager (*jicho*) or higher.

Although the ratio of white-collar to blue-collar employees is comparable in the auto industry of the two countries, differences appear when employees' occupational classes (e.g., technical or nontechnical) are also considered. A survey of vehicle manufacturers revealed that, in Japan, the proportion of engineers and technical support personnel reaches nearly 14 percent of total employment (7.6 percent engineers, 6.2 percent technical support). In the United States, the total engineering share is only 6.5 percent (4.1 percent engineers, 2.4 percent technical support).[39] The functional employment of engineers is, on average, about the same in both Japan and the United States among vehicle manufacturers and parts suppliers. In all cases, slightly over 50 percent of the engineers are in product engineering, and the rest are in manufacturing. In addition, some engineers may be assigned outside product or manufacturing engineering, such as those in marketing or corporate planning.[40]

However, notwithstanding the averages, large differences can be observed within the industry in each country. In Japan, most of the differences can be attributed to the company's age and size (the smaller and younger the company, the higher share of engineers). Similarly, in the United States, the larger the firm, the smaller its share of engineering personnel. In general, a higher ratio of engineering personnel may indicate a higher potential for new product development and improved process engineering. At the same time, the size of engineering overhead, if excessive, may negatively affect the competitive position of individual firms.

One could even argue that under some circumstances a lean engineering staff would give an advantage to the U.S. firms if the "savings" in the engineering manpower were not absorbed by expansion in administrative personnel. As a result, and as pointed out above, the final ratio of white-collar employees to the total labor force is similar among OEMs in both countries. The disparity in use of administrative personnel is most pronounced in the finance and accounting function. In American Company A, the number of salaried personnel employed in these two functions can be counted in the thousands, reaching 8 percent of total salaried employment; in Company B, the ratio is 10 percent. In contrast, in Japanese Firm X, it is less than 2 percent, and the total number of financial staff is well under 500. This is not to say that Japanese companies do not employ strict financial controls in their operations. However, the collection of financial data, as well as a substantial part of the financial analysis and planning, is delegated to line managers, and dual control systems are infrequent. Given the current average salaries for white-collar employees, the resulting cash savings per car are over $100.

It should be added that the composition of the white-collar labor force is today in a state of flux due to personnel cutbacks or freezes implemented in both countries during the world recession of the past several years. In U.S.

---

[39]In absolute numbers, however, over 50 percent more engineering and technical personnel are employed by the U.S.-based vehicle manufacturers, whose total employment is some 700,000, compared to 215,000 for the nine auto producers in Japan.

[40]Lawrence T. Harbeck, "Technical Manpower Characteristics of the U.S. and Japanese Automotive Industries," Research Report (Ann Arbor, MI: Joint U.S.-Japan Automotive Study, 1983), p. 7.

Company A, while the number of white-collar employees as a group declined in the same proportion as the number of blue-collar workers, white-collar employees assigned to sales decreased most, followed by technical support personnel, clericals, and managers and supervisors. In contrast, the number of technical professionals remained stable. As a result, their share of the total white-collar labor force increased by 10 percent. Similar proportional changes were observed in Company B, although, in comparison to blue-collar workers, the decline for the white-collar group was more than 50 percent less.

In Japan, the changing share of white-collar labor was due to differences in rates of employment increase, as the favorable competitive position of Japanese vehicle manufacturers protected their and their supplier's employees from an employment decline. However, a number of rationalization policies aimed at reducing the direct labor input were conducted throughout the period. As a result, the share of white-collar employment increased by 2 to 7 percent, depending on the firm. For example, in Company Y, since 1974, direct labor decreased marginally, indirect labor increased by 2.9 percent annually, administrative white-collar employees increased 4.1 percent annually, and the engineering professionals enjoyed the fastest growth, 4.7 percent per year.

Gender, Education, and Age Profiles

In both Japan and the United States, the proportion of females in white-collar jobs varies with occupational class. For example, in American Company A, women workers comprise nearly 60 percent of office and clerical employees, but only 18 percent of technicians and professionals and 6.5 percent of managers. It should be pointed out that, even though total employment in the U.S. automotive industry declined over the last several years, the proportion of women in technical and managerial jobs has increased. The same can be said about minority employment. In Company A, 9.5 percent of managers and 10 percent of technicians and professionals are minority employees.

In Japanese firms, 24 to 25 percent of white-collar employees are women, virtually all of them in nonmanagerial positions. Most of them are young office workers. For example, over 90 percent of women employed in Company X are younger than thirty. Although this may be changing, the vast majority of them still expect to retire at the time of marriage or soon thereafter and are seldom assigned to jobs that may lead to future managerial positions. This lack of promotional opportunities in return reinforces their motivation to resign. Thus, in both countries, white-collar employees in managerial or technical jobs are still predominantly male.

The educational profile of white-collar employees in Japanese and U.S. automotive OEM firms is also similar. The proportion of employees with a college degree ranges from 33 to 38 percent. Not surprisingly, the number of employees with advanced degrees is higher in the United States (6 to 10 percent), reflecting a higher reliance on formal training outside of the firm as opposed to on-the-job training practiced in Japan. Among managers, in American Company B, about 50 percent have a college degree, and 20 percent an advanced degree. Among technical employees, 35 percent have bachelors degrees and 12 percent an advanced degree. In Japanese Company X, only 3 percent of managers have advanced degrees, but 80 percent are college graduates. However, the proportion of engineers with graduate degrees has risen since the mid-1970s.

Among college-educated engineers, educational differences are greater between the U.S. OEMs and parts suppliers than between the U.S. and Japanese vehicle manufacturers. In Japan, 22 percent of OEMs' engineers have graduate degrees. This ratio is 19 percent for the U.S. vehicle manufacturers and 26 percent for the U.S. parts suppliers. Available data indicate no significant differences with respect to engineering disciplines. The mix is about 50-60 percent mechanical or related engineering, 13-20 percent electrical-electronic, 6-15 percent chemical-metallurgical, and 10-20 percent industrial and others.[41]

In general, the average age of white-collar employees is higher in the United States than in Japan: forty-one in contrast to thirty-three to thirty-four (thirty-six to thirty-seven for male employees only). The same is true of parts suppliers in the two countries. Not surprisingly, given the well-known low interfirm mobility among large Japanese firms, the average tenure is high for all Japanese vehicle manufacturers. However, long tenure is also characteristic for the white-collar employees in the U.S. auto industry. In Company A, only 5 percent of the managers have less than ten years of seniority, while 70 percent have been with the firm twenty years or longer. Only 15 percent are younger than forty years of age. Among engineers, 20 percent are younger than thirty, which is about equal to the proportion of those over fifty.

Among the top officers in American firms, over 80 percent have more than twenty years seniority with the firm, not much less than for Japanese vehicle manufacturers. It is interesting to note that among Japanese parts suppliers, the proportion of higher-level managers with seniority of less than ten years is higher for first-tier than second-tier suppliers: 47 versus 32 percent (Working Paper 13, p. 16). This can be explained by the fact that executive positions in first-tier suppliers are often staffed by retired officers from affiliated

---

[41]L. Harbeck, op. cit., p.11.

manufacturers. From the viewpoint of OEM managers, Japanese second-tier suppliers offer generally much less attractive retirement opportunities.

Organizational Hierarchy

Organizational hierarchy and the resulting span of control is best understood in terms of specific cases, as organization charts or position titles often do not present a true picture of the distribution of power and responsibility. For example, in the engineering area, U.S. vehicle manufacturers may use the same title for two positions that are not at the same supervisory level. These differences will perhaps diminish under current efforts in the United States to reduce levels of engineering management, which may lead to a standardization of titles, at least within each company. In addition, however, titles and levels of supervision do not match one another, and thus levels of supervisory responsibility are often difficult to decipher.

In basic organizational units, such as divisions or plants, the main differences in structure among the U.S. and Japanese vehicle manufacturers is in the span of control rather than in the degree of vertical differentiation. Generally, Japanese executives and managers are assisted by several deputies, who, while nominally second in authority, are not in the direct line of supervision. At the section level, the span of control of a typical Japanese manager is substantially wider. For example, in the engineering area, a Japanese engineering manager may supervise ten to fifteen employees as opposed to five to six employees in the U.S. firms. As a result, a Japanese equivalent of the American function of chief engineer or director may often supervise about three times as many employees as his U.S. counterpart.[42]

As pointed out above, significant attention is currently being given in the United States to reducing layers of management throughout the organization. In the factories, such efforts usually result in the elimination of general foremen or assistant plant managers. However, while such reductions are certainly useful in streamlining the organization, comparison with Japanese manufacturers reveal that a top-heavy structure hampers effective communication and decisions from the plant management upwards, rather than downwards. For example, in Company A, the chief executive officer is six or seven management layers away from a typical plant manager. In Company B, the picture is similar. In contrast, in Japan, plant managers are often appointed to the companies' boards of directors (though these are overwhelmingly "insider" boards) and are at most only two levels below the CEO.

Even adjusting for the relatively larger size of many Japanese manufacturing sites (with respect to the number of employees) in relation to the total size of the companies, it seems that in the United States vehicle manufacturers

---

[42]L. Harbeck, op. cit., p. 15.

may have considerable room for improvement in the efficiency of management at the higher corporate levels. Part of the "surplus" of administrative personnel discussed earlier is probably a result of "demand" for staff to assist numerous senior managers and executives on divisional and corporate levels. Also, it should be pointed out that Japanese vehicle manufacturers, while nominally smaller in size, in fact represent mainly the assembly and engineering part of a much larger corporate group. In that sense, the propensity of Japanese firms to organize a family of independent firms—in a legal sense—rather than to build single-firm corporate empires allows them time to reduce the levels of managers and free resources away from administration and control towards product and manufacturing development.

Personnel Control, Selection, and Development

In American auto firms, approximately 6-7 percent of white-collar employees are assigned to the personnel/industrial relations area. In general, the number of personnel/IR staff has increased in the past decade partly to handle tasks associated with new federal and state regulations, as well as to handle the expanded benefits. While the two functions are usually split in the corporate staff, they are often integrated at the division and plant level. The link between the personnel/IR functions and the CEO varies by company, but it is usually less direct than in Japan.

White-collar personnel control is mostly decentralized. For example, in Company B, the personnel staff of the head office (170 people) is responsible primarily for executive personnel control and general personnel policy planning; personnel control of managers just below the executive level is the responsibility of the functional staff; the remaining managers or supervisors and other white-collar employees are handled by division and plant personnel staffs. A typical divisional personnel office has approximately 25-35 employees, including a sizeable group administrating benefits; personnel staff in plants vary in size from 5 to 40 depending on the size of the plant work force.

The Japanese personnel staff is relatively more numerous than in American firms. For example, in Company X, 9.5 percent of all white-collar employees are assigned to the personnel division or to the personnel staff at the plant level. Nearly two-thirds of these work directly in one of the central personnel areas; the rest are guards, dormitory employees, and medical staff. Personnel control, as in most Japanese firms, is decentralized; the central personnel office has a staff of only 100 employees. It is responsible for personnel policies for both white-collar and blue-collar employees, although contract negotiations with the union are conducted through a specialized section. In a typical large manufacturing facility in Japan (with approximately 4,000-5,000 workers), the personnel department is staffed on average with 60 personnel specialists. In addition, 15 dormitory employees, 15 employees maintaining recreation facilities, and 10 on-the-plant medical personnel report to the head of the plant personnel.

One of the main tasks of the corporate personnel function in both countries is the hiring and training of white-collar employees. For employees with college educations, major auto companies in both countries rely on recruitment on college campuses. Exemptions exist (more frequently in the United States than in Japan), but the vast majority of new college-educated employees is hired straight from school or a few years after graduation.[43] In the United States, some new employees are also hired from graduate schools, a trend that is also increasing in Japan, at least for engineering personnel.

In both countries, recruitment planning for college graduates is coordinated by the central personnel staff. However, in Japan, all college-educated employees are usually recruited through the head office and then assigned to divisions and plants; in the United States, both direct and indirect placement is used. Also, in American firms, there is no difference in the recruiting process for engineers and other college-educated employees. Although direct walk-ins do occur, most prospective employees sign up for an interview on their college campuses. After three or four rounds of interviews, final offers are made. In Japan, administrative personnel are selected based on applications solicited through mass mailings and promotion efforts on campuses; engineers are selected using the university faculty or staff as an intermediary. In some instances, when demand exceeds supply, graduates are "oriented" in their selection of companies. Such a mechanism assures most firms a "fair share."

So far, there has been no general shortage of engineers in Japan, due mainly to past government efforts to expand the capacity of engineering departments. However, as the technological foundation of the automotive industry continues to change rapidly, many companies in the auto industry are facing the task of adapting to these changes by recruiting engineers with a particular technical knowledge. Those with electronic or information-processing backgrounds are in especially high demand. For the future, in view of the fact that Japanese companies will be required to rely more than in the past on their own "in-house" R&D, the demand for top-quality technical professionals is likely to increase. In order to respond to such a demand, a reform of the current system of university education may be essential.

In the United States, need for reforms may be more immediate as enrollment in engineering schools lags behind Japan and many other

---

[43]Until very recently, many college-educated employees of General Motors were graduates of the GM-sponsored "General Motors Institute." Direct institutional sponsorship was, however, recently discontinued.

developed countries. Two issues are of concern here. First, teaching jobs are unattractive relative to opportunities in the industry. Second, career opportunities in engineering fields are perceived as limited in comparison to law, finance, or management consulting. For example, among firms in the automotive sector, the percentage of engineers on the companies' boards of directors is over 50 percent in Japan, but only 20 percent in the United States.

To rectify the emerging imbalance, a joint effort of the U.S. corporate community and the public sector may be required, since problems in the area are beyond the control of any single company. However, in order to promote such an effort, better forward planning for engineering and technical manpower is desirable. Such planning is still in its infancy in the U.S. firms, as well as in Japan.

White-collar employees in both countries can take advantage of large training programs sponsored by their employers. The U.S. firms rely, to a greater extent, on courses and training offered by outside institutions, mostly nearby colleges. With few exceptions, such as senior manager-development programs at leading business schools, program, course, or seminar selection is left to the employee's initiative. For college courses, there are only very broad limits on tuition-refund programs. In Japan, in contrast, the emphasis is on internal training programs. Developmental planning is more structured, and course or program selection results from discussions between the employee and his/her manager and is monitored by personnel staff. On-the-job training is limited to lower-level, white-collar employees.

High-potential employees in both countries are often rotated through special developmental assignments. However, for the white-collar employees as a group, those in Japan have more opportunities to move for training purposes, although there is not much difference in the overall job mobility. For example, in American Company B and Japanese Company X, a significant difference in the overall volume of job mobility between the two firms was observed for technical managers only. For nontechnical managers, no significant difference in frequency of job changes was detected. In general, it seems that differences in mobility patterns are larger across the occupational class of white-collar employees (e.g., engineers versus nonengineers) than between countries.

In both countries, the functional job mobility of nontechnical managers is higher than the mobility of engineering managers and professionals. In other words, nonengineers have more general career experience than engineers. Also, the difference in functional specialization of technical professionals disappears when job functions are narrowly classified. Therefore, it seems that the popular perception that contrasts the Japanese "generalist" manager with the American "specialist" manager is, at best, an oversimplification.

Where the Americans and Japanese differ most is in the amount of interdivisional and interdepartmental mobility. Japanese managers, non-engineers even more than engineers, rotate through many parts of the organization (though often within the same function). This may increase their socialization into the firm, reduce costs of control and supervision, and improve communication and coordination. As a result, the system of job rotation in Japanese firms has a direct impact on their ability to facilitate organizational changes in general, and product development in particular.

Appraisal and Reward Systems

In most U.S. firms, the appraisal of white-collar employees is linked to an employee's potential. Management appraisals are coordinated by the central staff, but only higher-level managers and executives are reviewed by the executive office. Others are reviewed in divisions and plants. The process is annual and requires five months from the drafting of guidelines until the final review. The appraisal at all levels is performance oriented, although in several firms an evaluation of management style was recently introduced into the appraisal. Succession planning is an integrated part of the appraisal process, and promotions are seldom granted unless an appropriate position in the upper rank is available.

In Japanese firms, the appraisal system differs from the United States on several basic characteristics. First, a cohort of peers with similar education, seniority, and status forms the base for performance comparisons. Second, the evaluation process is centralized, and performance of all managerial-class employees is reviewed in the central personnel office. Third, an employee's performance is reviewed more often, generally at least two to three times per year, in conjunction with bonus payments in summer and winter and with salary and status reviews in spring or autumn. Finally, while self-report and an interview with the employee is one of the key components of the evaluation system, a decision to inform an employee on the appraisal results is left to the discretion of his or her immediate superior.

As for the actual level of compensation, comparisons between the United States and Japan are complicated by a whole range of factors, such as the cash value of benefits, or bonus eligibility.[44] Differences in career structures discussed earlier may also hinder the comparisons. However, at least partial comparisons are feasible if the focus of inquiry is limited to cash compensation payable to a majority of white-collar employees. In Japan, that would include a bonus that is payable to all employees, basically in proportion to their salaries. In the United States, bonus is excluded since over 80 percent of white-collar employees are not eligible for supplementary-compensation

---

[44]For discussion of the Japanese bonus and benefit systems, see Working Paper 3.

plans. Three questions are of interest here: the patterns of cash compensation in firms within each country, the absolute levels of compensation, and differences in compensation levels between white-collar and blue-collar employees.

At the entry level, the starting salaries in the United States are generally substantially above those in Japan. For example, for college-educated engineers starting salaries in the U.S. OEM firms range between $22,000-$27,000, in parts suppliers, between $20,000-$28,000. In Japan, starting salaries for engineers range from $9,500 to $10,500.[45] In both countries, the entry salaries are 15 to 25 percent above the entry wages for blue-collar personnel. However, as a consequence of the seniority wage system established in Japanese firms, annual wage increases granted to blue-collar workers follow the pattern set up for white-collar employees (in fact, the basic wage structure is the same for both groups). In the United States, blue-collar wages do not change after the first eighteen months of employment (except by across-the-board contractual increase or by reassignment to a skilled-trade job). A more detailed functioning of the two systems can be illustrated in the example of salary-grade distributions in cohorts of white-collar employees with twenty years of seniority in American Firm B and Japanese Firm X.

In the U.S. case, the reward system is structured along salary grades and ranks. Just over 50 percent of the cohort is in Rank A (plant superintendents, sales managers, supervisors, and product engineering) with an estimated average salary of $43,000; 30 percent are in Rank B (estimated average salary of $54,000); 15 percent in Rank C (estimated average salary of $76,000); and 3 percent in Rank D, which also includes all executive positions (estimated average of $111,000). The average cohort salary is about $54,000, nearly 30 percent over the Japanese average of $42,500. However, in contrast to relatively broad salary differentials in the U.S. firm, the highest-paid Japanese employees receive only $2,500 over their cohort averages, and only less than 10 percent of employees receive less than $40,000. In other words, the average cash compensation for Japanese white-collar employees with twenty years of experience is equal to the cash compensation for the bottom 50 percent of the U.S. employees with the same seniority (figure 9).

Finally, an "average" U.S. white-collar employee with twenty years of seniority receives nearly 120 percent more in annual cash compensation than a blue-collar coworker of similar age; for top performers among the white-collar group, the difference increases up to 500 percent. In Japan, the difference is less than 40 percent considering the average compensation levels

---

[45]L. Harbeck, op. cit., p. 12.

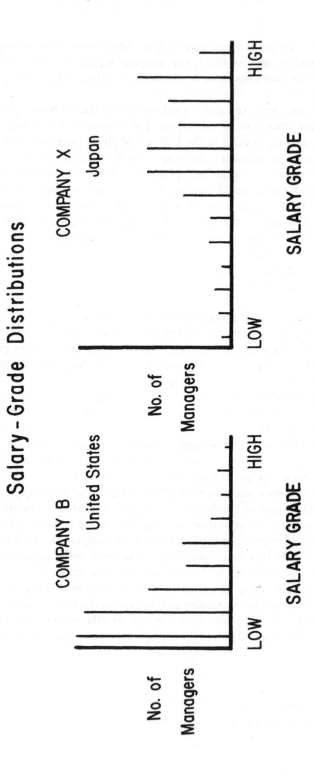

FIGURE 9
Salary - Grade Distributions

COMPANY B
United States

COMPANY X
Japan

No. of Managers

SALARY GRADE

LOW   HIGH

SALARY GRADE

LOW   HIGH

SOURCE: Cohort Salary Data—Company B and Company X.

for the two groups, and only marginally higher for the elite. Thus, the income differential between the two groups of employees is, on average, nearly three times as large as in the U.S. firm, nearly eight times as large in the case of the elite. This is a conservative estimate that does not include the bonus payable to the high-ranking American employees.

Similar trends become apparent when the salaries of key executives of the vehicle manufacturers in the two countries are considered. For top engineering officers, the range for the U.S. firm is between $100,000-$200,000.[46] In Japan, the range is $80,000-$90,000, even though their position in the corporate hierarchy is generally higher. Also, a Japanese CEO salary, including bonus, is generally not more than six to eight times larger than the income of the highest-paid blue-collar employee. For the United States, even if bonus and stock options are excluded, the CEO salary is usually twelve to eighteen times larger, again more than double the income differential observed in Japan. However, in profitable years, when American executives collect bonuses and stock options, this difference may triple.

Two explanations can be suggested for this phenomenon. On the one side, the lower income differentiation in Japan can be attributed to the deliberate policy of maintaining the cohesiveness of the organization by reducing salary differences across different strata of employees. On the other side, it can be argued that such a policy can be effective only when the dominance of internal labor markets in the economy restricts the mobility of movement across different companies or industries, especially for high-ranking managers and executives.

With respect to benefits, again, direct comparisons are difficult due to differences in accounting procedures (e.g., retirement payments). However, it can be estimated from the available data that in both countries retirement, health care, insurance, and other welfare benefits provide an additional 20 to 25 percent of compensation, not counting social-security taxes payable by the employers. It should be pointed out, however, that, in the case of Japan, a number of benefits are not costed out in wage or benefit statistics, the most important being housing mortgage subsidies.[47]

Future Trends and Policy Implications

In the future, the auto industries in both countries will have to adjust their human resource management systems to the profound organizational changes stemming from the accelerated introduction of new technology. For example, CAD and CAM in product engineering, the automation and

---

[46]Ibid.
[47]For a discussion of Japanese benefits, see Working Paper 3.

robotization of manufacturing processes in process engineering, and office automation in administration will necessitate changes in the employment structure of white-collar employees as well as require a massive retraining if the present employees are to qualify for new jobs.

Although future growth may vary by company, the maturation of the auto markets limits future opportunities for growth for the industry as a whole. This is compounded in Japan by the rapid aging of its labor force. By extension, opportunity for employee advancement may be restricted as well. If unchecked, this may result in a decline in motivation, especially among white-collar employees, and a loss of the talent necessary for these enterprises to other, still growing industries. The motivational "technology" may become an important strategic resource. Under the circumstances outlined above, the centrality of the human resource function to competitive strategy will increase dramatically. Future human resource management strategies in the white-collar area should foster flexible and timely adaptation to technological change, encourage innovation, and, most generally, provide equitable treatment and career opportunities to all employees in order to mobilize their creative potential.

The design and implementation of successful human resource management policies will, however, require strong and steadfast support from the top levels of management in each enterprise. In this context, while there are numerous differences in white-collar human resource management practice between the auto industries in the United States and Japan, there are also many similarities, and an exchange of experience in coping with some of the new challenges may prove to be mutually advantageous.

Chapter 12
Policy Implications

In concluding this report, the researchers were asked to develop a set of policy implications that arose from the research findings of the various areas under investigation. In addition, intensive discussions among researchers and with Policy Board members focused on our assessment of policy-relevant areas relating to the auto industry, even if they were not the subject of extended research on our part. Out of these considerations, we developed the policy implications to be discussed below. The broad headings of each section parallel the research topics covered in the report. It should be kept in mind by the reader that the Policy Board members themselves did not draft these implications; they are the product of the research efforts.

Trade Issues

Basic Understanding

The fundamental justification for free trade is that the resulting gains in efficiency enhance consumer welfare and the prosperity of the world economy as a whole. These principles were given expression by the commitment of the United States, Japan, and other world nations to GATT. Nevertheless, restraints have been practiced by governments in varying degrees through history. The political realities are that governments will resort to protection to overcome problems seen as threatening their national interest or the viability of major industries (including associated employment effects). The task, then, is one of guiding any resort to protection in a way that minimizes the problems posed for the world trading system and the costs to consumers and producers.

Observations

The liberal international trade regime is fragile because of the dilemma that it is in each country's individual interest to violate the rules as long as its trading partners do not. If all countries defect, however, the open regime collapses, and with it go the mutual gains from trade.

Nevertheless, the liberal trade order may result in very concentrated adjustment costs as particular industries are disadvantaged. Because the costs of protection are diffused while the benefits are concentrated and because sharply focused interest groups have disproportionate political power, governments have a tendency to react, even if in the aggregate costs exceed benefits.

There is no one cause of the current trade problems. Many factors have contributed, and the relative importance of each is uncertain. Furthermore, macroeconomic problems and the inherent problems of the auto industry itself must be clearly distinguished.

The policy instruments commonly discussed—tariffs, quotas, and local-content legislation—all represent what are, in essence, taxes on domestic consumers and subsidies to domestic producers. They differ primarily in their total cost, effectiveness, distribution of benefits, and political acceptability.

Policy Implications

Governments and public opinion must be reminded of the cost of protection to consumers and of its eroding effect on the liberal trade order. The problem of the U.S. auto industry is that so many short- and long-run factors all came together at once. As the economy of the United States and the world recovers, this will relieve some pressure for protection. Nevertheless, long-term problems will remain and will reassert themselves at a later date should underlying causes not be dealt with. Successful resolution of these issues is of obvious importance to producers in both countries.

The problems intrinsic to the industry must be attacked. However, it is also vital to remember that solving industry-specific problems may not be enough if the context remains one of an unfavorable economic milieu.

Any protection measures should contain an explicit schedule for their dismantlement in order to reduce uncertainty for the industry and to insure that the industry accepts the inevitability of the need to adjust. This is important since industry acceptance of this need to adjust provides an important incentive for management-labor cooperation. Such cooperation is indispensable for sustaining improvements in competitive standing.

Exchange Rates

Basic Understanding

Most observers believe that the dollar is overvalued vis-à-vis other currencies, and particularly the yen, in the trade context, though there is no evidence to support the view that the value of the yen is being manipulated. Exchange rates, however, have presented a fundamental problem for the U.S. auto industry and by and large benefited Japanese producers.

Observations

The *variability* of exchange rates makes planning difficult. While parity in purchasing power may hold in the long run (say twenty years), in the short run, deviations may be substantial and persist long enough to cause serious economic harm.

Even if a solution to the variability of the exchange rates is found, this does not mean that the resultant *level* will be satisfactory. It is possible that, because of the nature of the economies involved and the magnitude of financial flows, the exchange rate that produces equilibrium in these total flows may nevertheless pose problems for trade, and particularly trade in specific sectors.

The present exceptional strength of the dollar, relative to almost all other major world currencies, is heavily influenced by capital flows associated with America's high real interest rates and inflows of "safe-haven" funds. These factors appear to be thwarting the self-adjusting tendencies of free exchange rates, in which substantial merchandise trade imbalances tend to bring about changes in the currency rate that ameliorate the imbalances. It is not clear whether joint international management efforts to correct this situation are practical and workable, but unless and until these disequilibria are resolved, political pressures to support industrial bases threatened by sustained exchange-rate anomalies will clearly be intense.

The exchange-rate issue is not solely a bilateral one. Final products incorporate raw materials, components, and other inputs from a wide variety of countries, meaning that a number of bilateral exchange rates play some role in determining final prices.

Because long-run exchange-rate levels are the consequence of so many aspects of the relevant national economies, it is difficult for governments to deal effectively with the problem. National governments have been unwilling to sacrifice domestic policy goals to achieve improvements in exchange-rate levels.

Policy Implications

The governments of the major developed economies should be encouraged to coordinate their monetary and fiscal policies to avoid the overshooting phenomenon of changes in the exchange rate.

The Japanese government should be encouraged to continue its financial liberalization measures, including the freeing of interest rates from administrative control and the developing and freeing of Japanese capital markets for access by foreigners as well as Japanese institutions. These actions will contribute in the long run to increasing the value of the yen as it then becomes more of an internationally used currency.

Bilateral Market Access

Basic Understanding

Throughout the post-World War II period, the United States has had low tariff rates on passenger cars compared to other major trading nations, with the current tariff standing at 2.8 percent. Japan, historically, has had high tariff rates, but these rates fell rapidly in the 1970s, reaching a zero level in 1978. Even after the completion of the Japanese efforts at dismantling their nontariff trade barriers, the imbalance in respective market shares between Japan and the United States is not likely to be redressed in the near future.

Observations

The lower Japanese manufacturing costs plus the costs of transportation and associated preparation costs in Japan make extremely unlikely any significant penetration of the Japanese market by U.S. automotive exports. Furthermore, those size cars in which the Americans are preeminent are not suitable for the mass-volume market in Japan. These observations hold especially in the case of finished vehicles and, with some exceptions, tend to be the case with automotive parts and subassemblies as well.

This situation has been obscured by the slow process of dismantling nontariff barriers relative to Japan's world economic standing.

With the increased sourcing of finished vehicles and automotive parts from Japan by U.S. manufacturers, the imbalance in auto trade in Japan's favor could well grow.

Policy Implications

To avoid any charge of unfairness, the Japanese government should insure that all the last vestiges of protectionism in the auto sector are dismantled. This includes "leaning over backwards" to modify normal commercial practices to encourage imports. An extended emphasis from barrier reduction to import promotion would be welcome. For example, government action that would introduce foreign cars into government auto-procurement practices would be seen as an important symbol for such efforts.

Since the imbalance in auto trade is not likely to be redressed in the near future, it is important that Japanese government policy generally be oriented more strongly toward addressing the imbalance in their trade position. Sectoral imbalance in a particular area such as auto is less likely to become a serious economic or political problem if a better balance in the trade picture generally is achieved across sectors.

While Japan is not likely to become a major importer of finished vehicles, there is a potential for importing more components and raw materials. This outcome requires strenuous efforts on the part of the Japanese to overcome past barriers, as well as efforts on the part of potential foreign suppliers to meet Japanese specifications. American producers must realize that, as the Japanese government has gradually dropped import barriers, the responsibility for their success in exporting motor vehicles to Japan rests increasingly on them. Unless motor vehicles exported to Japan meet Japanese consumer standards of high quality, maintenance service, and cost and are promoted by effective dealer organizations, they will not make inroads on the Japanese domestic market. Moreover, just as Japanese producers adapted to U.S. marketing systems, the American producers must adapt to Japanese marketing systems to be successful. Even if American performance in these areas is outstanding, however, underlying cost and currency differentials make a major American auto-export program very unlikely.

Market Growth

Basic Understanding

Although growth in the overall world auto market has slowed, particular geographical markets and product areas show significant prospects for growth.

Observations

Due to slow income growth, weak economies, lack of industrial infra-structure, etc., the auto market in the Third World is not promising in the near future, though its potential in the long run cannot be doubted. Selected markets do offer opportunities for market growth.

The overall demand for automobiles in the aggregate is stable and depends on slowly moving, long-term influences. However, sales of new automobiles depend much more on current economic conditions. In particular, although the growth in aggregate demand in the United States and Japan is moderating, problems associated with cyclical instability will intensify in both countries. This instability has the potential to force production cutbacks, with resulting capacity underutilization and loss of employment.

The auto market in industrialized countries is becoming largely a replacement market. This leads to a more competitive market where product differentiation becomes an increasingly important corporate strategy. Under such conditions, high product quality and flexible manufacturing strategies provide important competitive advantages.

The Japanese carmakers today confront increasing difficulty in gaining access to foreign markets and are experiencing sluggish growth in their home market. This reflects, in part, inadequate growth in domestic demand for goods and services generally. In this context, the incentives for cooperation with foreign producers will continue to grow. At the same time, American producers under strong competitive pressures will continue to seek out cooperative ventures that reduce development, production, and marketing costs. The resultant international cooperation among producers will not necessarily take the form of merger since there are a variety of cooperative strategies by which companies can benefit.

Policy Implications

In the context of slowing aggregate growth in auto markets, only firms that can enhance their technical capabilities to meet these demanding market requirements may survive. Policies that would hamper firms meeting these market requirements should be minimized.

In recent years, Japanese cars have established, with regard to fit and finish, handling, and total car reliability, a firm reputation for better quality ratings than American cars. Even though American producers lead in some other areas, such as corrosion resistance and safety, the former factors tend to dominate consumer thinking on quality. American producers are making significant progress in improving overall product quality. If they are to be successful under the new market conditions described above, they must continue to make strenuous efforts in these quality-improvement activities.

Rapid and sudden changes in consumer preferences imply the need for corporate flexibility (e.g., flexible manufacturing systems). Those companies in a position to maintain alternatives in product offerings, markets, and production locations will have enhanced competitive strength.

The upturn in the U.S. economy has given domestic producers an opportunity to improve volume and profitability dramatically, especially under the lowered break-even points resulting from their cost-cutting efforts. It is critical that these efforts be continued. Without sustained further improvement, the evidence suggests that, in the absence of quota restrictions or with a subsequent economic downturn in the U.S. economy, there would be a significant decline in the market share for the domestic producers.

Strategic Implications of Technology

Basic Understanding

The complexity and sophistication of automotive technology is increasing rapidly. This trend will continue through the forseeable future and will lead to continued detailed but important changes in the product and production process.

Observations

While the automotive market in Japan, the United States, and Western Europe is becoming more mature, this is certainly not the case for automotive technology. In fact, a mature market should increase the development and application of new technology, which will lead to improved overall vehicle performance and value as the manufacturers and their suppliers attempt to improve market share and create new markets. Combined with the internationalization of markets, this will lead to the competitive position of auto companies being established on the basis of technical capabilities.

From a macro view, the vehicle is evolving in such a way that few fundamental changes will be evident to the consumer. However, on a micro scale—that is, within various vehicle systems and subsystems—a host of revolutions are in progress. The application of advanced electronic and communication technology, new materials, advanced engines and transmissions, and improved processing technology are but a few areas of fast-moving technology. Progress in all vehicle-related areas will be accelerated by developments in the traditional supplier industries and industries peripheral to the automotive industry. Furthermore, with trends to international sourcing and reduced vertical integration on the part of the U.S. manufacturers, technology will be increasingly internationalized. For the Japanese, the need to locate facilities abroad and to find foreign partners will facilitate the internationalization of technology.

Important changes within the traditional supplier industry are expected, particularly in the United States because of the fast pace of downsizing. However, the trend to front-drive vehicle reconfiguration, lighter materials, and electronics in both the United States and Japan suggests that supplier dislocations are likely in both countries.

Fuel economy will undoubtedly remain as one of the most important driving forces in the development of automotive products. However, there are significant differences between the U.S. and Japanese markets. In Japan, fuel prices are expected to remain high and be substantially greater than U.S. prices. In the United States, industry experts believe energy prices will increase only slightly from present levels and remain far below world standards. Consequently, they expect continued differences in vehicle consumer requirements in the United States and Japan, which will be reflected in numerous technological factors.

Another point of fact in the international industry is the significance of the industry as a user of scarce and expensive materials and energy. In many cases, the industry is the dominant user, for example, of rubber, steel, and specialty items like platinum and rhodium.

Policy Implications

With the growing importance of technology, policy actions that will encourage company R&D activities is recommended. This applies to both proprietary R&D within a given company and support of joint, basic or fundamental, and nonproprietary research efforts.

All companies and organizations influenced by the automotive industry should carefully monitor domestic and international technical trends in order to assess the impact of change. This may require a defensive strategy and a shift to other opportunities or an offensive strategy to capture developing markets.

Despite markedly lower expectations for the rise in energy prices in the United States than just two years ago, little change in vehicle technology is forecast from a period when expectations were for a much higher rise. Undoubtedly, some skewing of vehicle model mix will occur, but U.S. manufacturers will not retool to build much larger cars. Thus, while energy prices may fluctuate, vehicle trends will hold on a steady course of downsizing and weight reduction, although not at the rate originally predicted.

In Japan, energy prices are expected to track with normal inflationary movements but to remain much higher than U.S. prices. The average car sold in Japan will continue to be considerably smaller and lighter than those sold in the United States.

Investment and Technology

Basic Understanding

Investment requirements by the auto industry will continue to be higher than historical levels.

Observations

Technological advances in both product and manufacturing are more rapid than in the past.

Competition is increasing as national markets become international.

Presently unforeseen technological changes could accelerate already high levels of change.

Policy Implications

All firms should plan for continuing high levels of capital need.

Weaker firms should focus and concentrate their capital employment to ensure technical parity in chosen market segments.

Any government action that would encourage investment is highly desirable in order to hasten technological changes.

Government policy that would excessively force technology should be avoided. Furthermore, consistency and a long-term focus of regulatory policy is necessary to avoid distorting company product-development efforts and, therefore, investment demands.

Manufacturing Cost Differences and Productivity

Basic Understanding

The U.S. auto industry currently faces a real and substantial disadvantage in manufacturing costs compared to the Japanese industry. The significance of this difference is influenced by many factors, but it is clear that it constitutes a long-term problem that must be addressed.

Observations

The size of the reported manufacturing cost difference (MCD) depends on a number of factors that are treated differently in the various public reports. These include the yen-dollar exchange rate, the method of adjusting for different product mixes and differing degrees of vertical integration, the inclusion of different sources of the MCD, and the rates of the industry's capacity utilization. Nevertheless, despite the incomplete and conflicting nature of these studies, it is clear that there is a very substantial MCD for small cars.

The MCD is quite volatile.

The greater the market share of small, fuel-efficient vehicles, the greater the significance of the MCD to the overall competitive performance of the U.S. industry.

Comparably equipped U.S. and Japanese passenger cars are being priced quite similarly in the U.S. domestic market. The typical publicly available estimates of a minimum of $1,500 per vehicle MCD imply that the Japanese producers and their U.S. dealers selling in the United States are making substantial profits, while the U.S. industry is incurring losses in the small-car market. The U.S. industry's profits from larger cars are, in effect, subsidizing their participation in the small-car market, while the Japanese North American profits, in effect, become available for pricing strategies in other markets.

The U.S. industry needs to identify areas of potential cost reduction through analysis of its own situation. This must extend beyond comparing its cost structure to the Japanese industry and into factors external to the industry, as well as compensation systems, costs of medical care, etc.

The U.S. industry needs to pay careful attention to the long-term nature of the MCD. In particular, the necessary short-term responses must be evaluated in terms of their implications for long-term strategies.

Policy Implications

The U.S. industry must address the MCD in a concerted, consistent fashion and not pursue short-term actions that close off or undermine necessary long-term strategies. They have already made significant progress, as indicated in significantly reduced break-even points. Historical experience, however, indicates that a return to financial health often leads to a return to bad habits. A way must be found to address the MCD over the long term.

U.S. government policy makers need to recognize that large fluctuations in exchange rates can offset steady progress in reducing cost differences. To some extent, this has occurred between 1982-83.

Some researchers believe that the Japanese companies need to be cautious about adopting a price-competitive strategy within the U.S. market. Pricing that reflects a level near the reported MCD, coupled with a large sales increase to meet all demands at that pricing level, would surely lead to strong political pressure on the U.S. and Canadian governments to further restrict Japanese access to North American markets. Other researchers stress that this restraint constitutes a restrictive practice that would sacrifice consumer interests. Thus, the only solution must involve continued intensive efforts to improve the competitive standing of U.S. automotive firms.

Sources of Manufacturing Cost Differences and Productivity

Basic Understanding

The publicly available information about the *sources* of the manufacturing cost difference (MCD) are less clear than they should be if effective long-term actions by the industry are to be undertaken. This is because the optimum strategies for responding to the MCD will vary depending on the weight one attaches to different sources of the differential.

Observations

The role of many factors—material costs, transportation charges, inventory practices, quality-assurance practices, capital costs, taxation policies, to name some of the key factors—have been inadequately analyzed in the publicly available materials.

Most analysts agree that "productivity differences" and "wage-rate and benefit differences" are the two major factors accounting for the MCD, although they disagree as to their importance relative to each other.

If the bulk of the differential is due to wage rates, then strategies designed only to allow time to "learn from the Japanese" are not likely to succeed, whether at the government or industry level. Under such conditions, those strategies likely to be more effective are: adopting labor-saving technology, moving operations to areas with lower-cost labor, or overhauling U.S. compensation practices, with greater emphasis on profit sharing and other offsetting benefits and less on fixed wages and benefits. (Obviously these problems would be ameliorated if labor costs per man hour in the major producing countries were to be made equal.)

The increasing age of the Japanese auto worker, as well as increases in real income, is quite likely to raise Japanese wages over the next decade. This is likely to decrease the MCD vis-à-vis the North American industry and to make Korea, Taiwan, and other new entries more of a worldwide competitive threat to Japan.

If the bulk of the differential is due to productivity, then "breathing space" strategies make more sense. Under these conditions, strategies to benefit from lower labor costs may exacerbate productivity disadvantages and, therefore, may not be as effective long term.

On balance, productivity sources appear to be more important, and probably more correctable.

Policy Implications

Industry participants need to carefully tie strategic responses to the MCD to well-documented sources.

For reducing MCD, it becomes of vital importance to make every effort to improve labor-management practices and to promote quality assurance and manufacturing technology. American auto firms have already made considerable headway in upgrading quality and reducing production costs. With a continuation of the upturn in the economy, these efforts will bear more fruit.

Since there are built-in institutional rigidities limiting the amount of further progress that can be made in narrowing compensation differences (regardless of the extent to which compensation differences contribute to the MCD), it is all the more important for U.S. parties to work on eliminating productivity differences.

The Role of Suppliers

Basic Understanding

The contributions of suppliers play a decisive role in the competitive situation of their customers. Their flexibility in response to changes in the strategy of auto manufacturers is critical.

Observations

There is an increasing interest on the part of American auto manufacturers to benefit from the technology and productivity of Japanese suppliers.

There is a parallel increased interest in joint cooperation between Japanese and American supplier firms.

Thus, international trade in auto parts, technology, and equity flow is increasingly significant.

American suppliers desire to strengthen their technological contributions to customers both individually and through multi-supplier joint efforts. The lack of joint test and development facilities (analogous to the Japan Automotive Research Institute [JARI]) in the United States may adversely affect these joint efforts. Confusion regarding antitrust implications may also hinder these actions.

Policy Implications

Mutually beneficial international trade in auto parts, technology, and capital flows should be facilitated. Trade flows in auto parts in Japan's favor must be evaluated in terms of possible offsetting trends in technology and capital movements. A close monitoring of these various trends is necessary to grasp the full picture of the respective benefits of automotive trade.

Changes in applicable laws and regulations necessary for effective implementation of joint development programs should be seriously considered by appropriate U.S. legislative and executive bodies.

The Changing Structure of the
Manufacturer-Supplier Relationship

Basic Understanding

Although the formal structures and practices for manufacturer-supplier relations are now quite different in the two countries, they will probably become more similar over the coming years.

Observations

Although we will see some changes from both sides, more of the relative movement will come from the American side. There will almost certainly be a more compact and hierarchical supplier base in the United States after this period of transition, perhaps more analogous to that in Japan. Much of the motivation for change on the Japanese side will come from the needs imposed by the facts that automobile manufacturers are becoming more international in their manufacturing base and that suppliers are attempting to establish greater independence of action.

If the American automotive companies put into effective practice the programs they have announced, they will be well on the way to accomplishing much of what is necessary with respect to improving the joint technological innovation and productivity improvement of themselves and their suppliers. Many suppliers are concerned about the manufacturers' long-term commitment to institute the proposed changes.

The effectiveness with which U.S. industry creates new price-setting mechanisms between manufacturers and suppliers will be an important element in their efforts to establish a new basis of cooperation.

Policy Implications

American producers must develop more cost-effective component designs through cooperative manufacturer-supplier efforts in order to become more competitive in productivity and quality. Important to the success of this effort is the development of new approaches to setting vendor prices and transfer prices (those between one division and another).

Organizational development, management selection, and training in automotive purchasing departments must be adapted to the structures and personnel necessary to implement these programs effectively. In the United States, this will require an influx into purchasing of staff with certain technical (in both the engineering and other senses of the term) skills and experience. This will require significant cross-functional training and other changes in career paths. Supplier organizations and staffing must adapt to the new requirements implied by these new approaches.

Japanese auto manufacturers should continue and intensify their efforts to obtain the services of new suppliers (including non-Japanese firms) who provide better cost, quality, and/or delivery. Organizations and mechanisms should be adapted to facilitate new and more internationally based suppliers.

Human Resource Development

Basic Understanding

Human resource management is a major factor in competitive strength.

Observations

Since the 1930s, industrial relations in the U.S. auto industry evolved around a system of collective bargaining. With this system, the union had some success in minimizing the discretionary powers of employers (e.g., determining wage rates, job allocation, work loads, etc.). This reduction in arbitrary decision making led workers to view collective agreements as the primary means to protect worker interests against management power that controls production, allocation of capital resources, and choice of technology. American automotive management has not readily shared information with the union or the workers and has been relatively disinterested in the ideas of the labor force. As long as the U.S. auto industry was enjoying an overwhelming technological and market advantage, this system of industrial relations was compatible with high productivity and wages.

Industrial relations in the Japanese auto industry also have developed around a system of collective bargaining, one transplanted from America. Unique features developed in Japan, including a bargaining structure that is confined to each enterprise rather than being industry-wide and a system of joint consultation between management and labor. These features facilitate information sharing on both sides. Functionally, Japanese manufacturers stress in-company training, both off and on the job. Compared to the United States, Japanese collective agreements are less rigid in prescribing working conditions. The union and the workers exert a considerable influence on a wide range of issues through workshop consultations and small-group activities. These issues include even choice of technology and the mode of its implementation.

In recent years, the emergence of adverse economic conditions in the U.S. auto industry has fostered a recognition on the part of both management and labor of the limitations of conventional industrial relations. What has been particularly noted is a lack of flexibility, of worker input, and of sharing management information with the work force. Efforts to redress these issues are now being pursued at many U.S. plants, with differing degrees of intensity. Management seeks to gain flexibility by modifying work rules and to increase productivity and product quality by mobilizing new techniques of labor management. The union accepts wage moderation and changes in work rules and demands a greater degree of job security. Activities such as QWL campaigns, EI programs, QC circles, and joint consultation are still in the early stages of development. These measures are indispensable for industry survival and worker security in the face of ever accelerating technological change and the changing industrial structure.

In Japan, too, the basic factors supporting the vigorous performance of the auto industry have been changing substantially in recent years; the labor force is aging rapidly, the growth prospects of the market are diminishing, the need for internationalization through direct foreign investments is growing, and new waves of technological innovation are influencing both working conditions and employment while suggesting new product frontiers. Japanese management is faced with a serious challenge as to whether it can adapt to major structural changes.

Policy Implications

Under these circumstances, a management strategy that will be the most important both for the U.S. and Japanese auto industries is systematic and effective human resource development. Fostering a productive, creative, and adaptable work force will be the single most important measure for assuring the survival and facilitating further development of the industry in an era of rapid technological and structural changes.

One of the crucial conditions determining the flexibility or inflexibility of management and labor in coping with changing environmental conditions is the extent to which both parties understand and share information about the situation. Joint problem solving needs to be pursued to tackle the specific issues faced by both the U.S. and Japanese auto industries.

Technical Manpower

Basic Understanding

Both the U.S. and Japanese automotive industries are increasingly based on the application of high technology. This includes product, manufacturing, and service aspects. In many areas, the automotive industry is in a position of technical leadership, such as the application and development of CAD/CAM technology. This increased technological sophistication and complexity requires a large number of outstanding technical employees (engineers and engineering support personnel) to maintain and enhance the competitive position of both automotive manufacturers and suppliers. The competitive challenge exists between companies within a given country and between countries.

Observations

Manpower needs are determined by product, process, and service requirements of current and future vehicles. These needs are only in part effective in influencing the availability of technical personnel. Organizations and institutions affected directly by this demand are the vehicle-manufacturing companies and their suppliers and schools (including primary and secondary), colleges, and universities.

In both the United States and Japan, there is growing concern over the attractiveness of the automotive industry to engineering graduates. In the United States, the concern is with both the number and quality of graduates, whereas in Japan, the principal concern is with quality.

Automotive technical manpower needs are a complex mix of highly trained persons in the several engineering and scientific disciplines, including computer science.

Industry support of education is largely indirect through product sales, jobs, taxes, and tuition growing out of the production process.

Continuing education is required for all technical personnel to insure performance at the "state-of-the-art" of technology. Relatively speaking, more of this continuing education is supplied within the firm in Japan as compared to the United States.

Policy Implications

The automotive industry should develop longer-term plans for technical manpower needs and make their forecasts available to the educational system.

Financial support sufficient to train, update, and hire faculty and modernize facilities, including laboratories, must be expanded in universities, companies, and training institutes to produce the needed technical personnel. This is especially important in the United States.

Expanded and systematic training programs should be developed to augment modest, current, formalized efforts and "on-the-job" programs. These should be directed at all technical levels in a given organization.

Industry support for universities should be expanded to support the automotive industry's technical manpower needs. In large part, this support could take the form of sponsoring directed, basic research.

The automotive industry should further strengthen R&D programs, and these should be well publicized to insure that students and faculty become aware of the needs and exciting opportunities in the industry.

Public Policy

Basic Understanding

Public policy, though important, has not been the major determinant of the successes and problems in either the U.S. or Japanese auto industries. Moreover, contrary to frequently encountered generalizations, in both countries governmental policy has had mixed effects, helpful to the development of the auto industry in some areas and periods, a hindrance in others.

Observations

Japanese industrial policies have been adjusted according to changing economic conditions. This implies that the policies at the stages of economic reconstruction (infant industries), rapid economic growth, and long-term uncertainty should be different.

The Japanese government policy with the greatest impact on the auto industry was the protectionism of the 1950s and 1960s, part of an overall protectionist orientation. However, beginning in the 1960s, there was a rapid reduction and eventual elimination of tariff barriers. Along with other manufacturing industries, the auto industry has also benefited from general economic policies in such areas as savings and investment, credit, technology imports, antitrust, and export promotion. Programs focusing on the auto industry, such as loans, tax concessions, and so forth, have been much less important, particularly in recent years when the relationship has been more distant and many government actions (environmental regulations, export restrictions) have had negative impacts.

Until the 1960s and again in the 1980s, the American government pursued various policies, some inadvertently, that aided the auto industry. However, in the critical period when import competition was gaining strength, energy and regulatory policies had a substantial negative impact. Inconsistency among policies and rapid changes of policy have been particular problems.

Policy Implications

The lesson Americans should draw from Japanese industrial policy, at least for the automotive industry, is not one of top-down planning and coercive implementation, nor one of heavy subsidies. The key elements are a careful balance between a nurturing environment provided by the government and the independence of the companies to compete and produce a better product for the customer, on-going government and industry communication, discussion about current trends and future problems, plus continued attention to the issue of international competitiveness.

In the case of the United States, governmental efforts to improve the international competitiveness of American industry are warranted. International competitiveness cannot be continually subordinated to domestic policy or defense policy without serious long-term damage being inflicted on the industry. At a minimum, the government must assess the impact of policy changes in other areas on the international competitiveness of the auto industry.

A fundamental question facing the American auto industry is whether the domestic market will continue the trend of the 1970s toward integration with the world market or return somewhat to a more "insulated" market. Government will not determine this choice, but public policy will influence it: for example, the proposal to raise gasoline taxes substantially leads in the former direction, and repealing CAFE regulations the latter. Implications of such choices must be thoroughly examined.

A highly complex pattern appears to be emerging of increased cross-national, intercorporate linkages of licensing, coproduction, purchasing, sales agreements, and so forth. Although neither this future pattern nor its implications are clear, it is likely that governments will need to play a role in encouraging, inhibiting, or shaping these trends as they affect such public concerns as competition, employment security, international relations, and so forth. Innovative approaches to policies in such areas as regulation, trade, antitrust, and taxation require imaginative consideration.

The question of whether "industrial policy"—targetted government assistance in access to capital, joint research, and so forth—is needed for American industry in general or autos in particular is currently much debated. In any case, government should act to aid reemployment and to redevelop regions dependent on auto production since the bulk of currently unemployed auto workers are unlikely to be rehired, even with economic improvement. For example, trade-adjustment assistance would be more effective if it were used for sustained financing of the delivery and actual retraining of workers, not simply for augmenting unemployment assistance. Both Europe and Japan provide several models of public-private cooperation in such programs that merit serious investigation.

During the period of rapid growth of the Japanese auto industry, the government did not play a predominant role in policies to promote growth *per se,* although its policies to deter the entry of foreign vehicles and capital were very significant. However, in the current period of domestic saturation and difficulties of increasing exports, auto sales are unlikely to resume their earlier growth rate, and a more careful government policy may be required. While we presume that the Japanese auto industry will not fall into the "structurally depressed" category in the near future, the Japanese government should engage in "future-oriented thinking" about these problems to avoid overly hasty reactions to emerging problems.

Appendix A
Working Paper Series

1. Cole, Robert E., and Hervey, Richard P. *Internationalization of the Auto Industry: Its Meaning and Significance.* 94 pp. October 1983.
2. Bittlingmayer, George. *The Market for Automobiles and the Future of the Automobile Industry.* 70 pp. July 1983.
3. Flynn, Michael S. *Differentials in Vehicles' Landed Costs: Japanese Vehicles in the U.S. Marketplace.* 65 pp. October 1982.
4. Cole, David E., and Harbeck, Lawrence T. *Evolving Automotive Technology.* 67 pp. March 1982.
5. Hervey, Richard P. *Preliminary Observations on Manufacturer-Supplier Relations in the Japanese Auto Industry.* 33 pp. June 1982.
6. Harbeck, Lawrence T. *The World Car.* 9 pp. February 1982.
7. Bittlingmayer, George. *Auto Scrappage and New Car Demand.* 20 pp. August 1982.
8. Iguchi, Masakazu, Kodama, Fumio, and Yakushiji, Taizo. *Comparative Delphi Studies of U.S. and Japanese Future Automotive Technologies.* 68 pp. August 1982.
9. Tschoegl, Adrian E. *The Yen: Villain or Not?* 21 pp. April 1983.
10. Kamath, Rajan, and Wilson, Richard C. *Characteristics of the U.S. Automotive Supplier Industry.* 49 pp. November 1983.
11. Cole, Robert E. *Access to Auto Markets: U.S.-Japan Bilateral Trade Relations.* 40 pp. August 1983.
12. Wilson, Richard C. *Future Automotive Factories: Speculative Models.* 26 pp. March 1983.
13. Kodama, Fumio, Yakushiji, Taizo, and Hanaeda, Mieko. *Structural Characteristics of the Japanese Automotive Supplier Industry.* 32 pp. June 1983.
14. Bittlingmayer, George. *World Auto Demand.* 9 pp. August 1983.
15. Liker, Jeffrey, and Wilson, Richard C. *Design and Manufacture of Automotive Components: Two Case Studies of U.S. Practice.* 80 pp. December 1983.
16. Campbell, John. *The Automobile Industry and Public Policy.* 33 pp. August 1983.
17. Iguchi, Masakazu, Kodama, Fumio, and Yakushiji, Taizo. *Second Comparative Delphi Studies of U.S. and Japanese Future Automotive Market and Production Technology.* 28 pp. December 1983.

18. Shimada, Haruo. *Human Resource Management and Industrial Relations in the Japanese Automotive Industry.* 62 pp. July 1983.
19. Flynn, Michael S. *A Note on the Treatment of Labor Content as a Source of the Manufacturing Cost Differential.* 17 pp. September 1983.
20. Flynn, Michael S. *Compensation Levels and Systems: Implications for Organizational Competitiveness in the U.S. and Japanese Automotive Industries.* 19 pp. September 1983.
21. Flynn, Michael S. *Estimating Comparative Compensation Costs and Their Contribution to the Manufacturing Cost Difference.* 28 pp. January 1984.
22. Cole, Robert E. *Information Systems and Competitive Economic Performance: Reflections on U.S. and Japanese Practices.* 15 pp. March 1984.

These papers are available for purchase from the Center for Japanese Studies Publications, 108 Lane Hall, The University of Michigan, Ann Arbor, MI 48109.

Appendix B
List of Project Sponsors

The Joint U.S.-Japan Automotive Study, conducted by Technova Inc., Tokyo, Japan, and the Center for Japanese Studies, The University of Michigan, Ann Arbor, Michigan, U.S.A., has been supported by:

Technova Inc.
Funds raised from:
Aisin Seiki Company, Ltd.
Nissan Motor Company, Ltd.
Toyota Motor Corporation

Center for Japanese Studies,
The University of Michigan
Funds raised from:
AMP, Incorporated
A.O. Smith
Barnes Group Inc.
Bendix Corporation
Borg-Warner Corporation
Government of Canada
Champion Spark Plug Company
Chrysler Corporation
Diamond Shamrock Corporation
Dilesco Corporation
Donnelly Mirrors
Earhart Foundation
Exxon Chemical Americas
Ford Motor Company
General Motors Corporation

Gulf + Western Manufacturing
Hoover Universal, Inc.
The Johnson Rubber Company
Kraftube, Inc.
State of Michigan, Department of
    Commerce
Motorola Inc.
Owens-Corning Fiberglas
Randall-Textron
R.J. Tower Corporation
Rockwell International
Saab-Scania AB
Schwitzer, Wallace-Murray
    Corporation
Spun Steel
Standard Oil of California
Stant Inc.
Technova Inc.
Tenneco, Inc.
Union Carbide
United Automobile Workers
U.S. Department of
    Transportation, Transportation
    Systems Center
Volvo Car Corporation

Printed and bound by CPI Group (UK) Ltd, Croydon, CR0 4YY

13/04/2025

14656537-0002